Force and Motion

FORCE AND MOTION

An Illustrated Guide
to Newton's Laws

JASON ZIMBA

THE JOHNS HOPKINS UNIVERSITY PRESS
BALTIMORE

© 2009 The Johns Hopkins University Press
All rights reserved. Published 2009
Printed in the United States of America on acid-free paper
9 8 7 6 5 4 3 2 1

The Johns Hopkins University Press
2715 North Charles Street
Baltimore, Maryland 21218-4363
www.press.jhu.edu

Library of Congress Cataloging-in-Publication Data

Zimba, Jason, 1969–
    Force and motion : an illustrated guide to Newton's laws / Jason Zimba.
      p.   cm.
    Includes bibliographical references and index.
    ISBN-13: 978-0-8018-9159-5 (hardcover : alk. paper)
    ISBN-10: 0-8018-9159-0 (hardcover : alk. paper)
    ISBN-13: 978-0-8018-9160-1 (pbk. : alk. paper)
    ISBN-10: 0-8018-9160-4 (pbk. : alk. paper)
    1. Motion.   2. Dynamics.   3. Mechanics.   I. Title.
QC133.Z56   2009
531′.11—dc22   2008033841

A catalog record for this book is available from the British Library.

Special discounts are available for bulk purchases of this book. For more information, please contact Special Sales at 410-516-6936 or specialsales@press.jhu.edu.

The Johns Hopkins University Press uses environmentally friendly book materials, including recycled text paper that is composed of at least 30 percent post-consumer waste, whenever possible. All of our book papers are acid-free, and our jackets and covers are printed on paper with recycled content.

Contents

Preface

Newtonian mechanics, the subject of this book, is no longer considered a fundamental theory of nature. We live in a world of quantum theory and nanotechnology. But ask a physicist of today, even a quantum physicist, to explain how a curveball works, and he or she will certainly use the methods and concepts of Newtonian mechanics to do so. Newtonian mechanics has survived its demise. It is still the most powerful set of ideas we have for explaining the phenomena of everyday life: walking, dancing, pushing a grocery cart, driving a car, flying in a plane. This is why the subject is still taught in schools and universities and why scientists still apply Newtonian mechanics in their research—more than 300 years after Newton first announced his Laws of Motion and more than 100 years after Einstein announced that he'd overthrown them.

Force and Motion contains a wealth of original problems that are carefully designed to build your understanding of Newton's Laws. I have been developing and refining these problems for over a decade while teaching hundreds of physics majors, engineering majors, premedical students, liberal arts students, and high school students from schools such as the University of California at Berkeley, Grinnell College, Bennington College, and Oakland High School in California. I developed these problems to address common difficulties I observed among my students that are not addressed effectively in any major textbooks.

The book is divided into seventeen chapters. Each chapter focuses on a single manageable topic, presented in a clear and approachable way, with an emphasis on what you really have to know and where people often stumble. The chapters include Worked Examples with detailed solutions and, at the end, Focused Problems to help you build your understanding of the topic at hand.

! When you finish reading a chapter, your work on that topic has just begun. Your next task is to solve the Focused Problems at the end of the chapter. Reading the chapters isn't enough. *Understanding* the chapters isn't enough. *Only by trying the Focused Problems* will you find out whether you can apply the ideas yourself.

Acknowledgments

During the writing of this book, I was blessed with the support, encouragement, and advice of many students, colleagues, friends, and family members. In particular, I thank Roger Berry, Bruce Birkett, Andrew Charman, David Coleman, Jane Hoehner, Charles Nix, Sonia Perez, Susan Sgorbati, Ann Shaw, Charles Wohl, and all of the students in my Introductory Physics classes at Bennington College.

It has been a pleasure working with Trevor Lipscombe and Bronwyn Madeo at the Johns Hopkins University Press, and I thank them for supporting this project and bringing this book to publication.

Nino Mendolia and Charlotte Welch of Bennington College gave me valuable assistance with some technical aspects of producing the manuscript.

The United Nations Environment Program, Information Unit for Conventions, kindly gave permission to use Figure 1 from the 2006 report of the Intergovernmental Panel on Climate Change.

I give special thanks to my wife, Rebecca. She made room for this project during a busy time in our lives, and she also read extensive portions of the manuscript, correcting many errors and improving the clarity of the exposition in a number of places.

Needless to say, any errors and obscurities that may remain are entirely my own.

INDEX OF KEY MATERIAL

PART I DESCRIBING MOTION

The ancient Greeks understood the word *motion* to mean not just the movement of an object from one place to another but, more generally, any change of a quantity over time, such as the changing water level in a lake or the changing size of a spreading stain.

Before you can *explain* why motion happens or *predict* what motions will occur, you must first be able to analyze *motion itself.* You must be able to understand and visualize sophisticated scenarios such as "The water is rising at a decreasing rate" or "Inflation increased at a rate that increased at a constant rate." And you must learn how to use the terms *position, velocity,* and *acceleration* in precise and rigidly defined ways. The aim of this part of the book is to help you understand these concepts.

CHAPTER 1

Graphing Relationships

Being comfortable with graphs is a basic requirement for citizenship in modern society. To become knowledgeable about issues such as global climate change, for example, we have to know how to interpret graphs like the one shown in Figure 1. This figure is a *line graph*. Line graphs are the most important kinds of graphs for understanding motion, because they illustrate how quantities change with time.

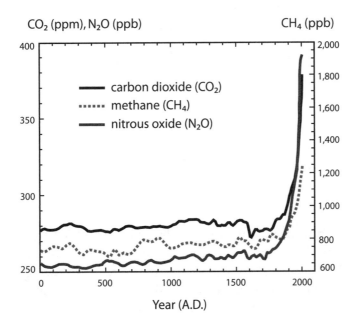

FIGURE 1.
A line graph showing carbon dioxide concentrations in the atmosphere over the past 2,000 years. Reproduced from Reference [1] with permission from the United Nations Environment Programme, Information Unit for Conventions.

Whenever you see a graph, take a moment to study the title, the horizontal axis, and the vertical axis. What quantities are being plotted? What are their units? Where is zero on the vertical scale? Are both positive and negative values present along the vertical axis?

The goal of this chapter is not to demonstrate how to graph a mathematical function such as $y = x^2$. Rather, I want to give you some opportunities to practice a much higher-level skill: sketching graphs to reflect an intuitive understanding of a relationship.

» WORKED EXAMPLE 1

Sketch a non-numerical graph of the height of a tree versus time over the lifetime of the tree. (I'm not asking you to look up the facts of tree growth for this problem; I just want you to graph the relationship as you think it might actually play out in nature.) On your graph, identify the periods of time, if any, during which the height is increasing. Also identify the periods of time, if any, during which the height is decreasing.

» SOLUTION 1

My graph is shown in Figure 2. Your graph may look different. (You did draw a graph, didn't you?)

FIGURE 2.
Solution to Worked Example 1. The height of a tree versus time.

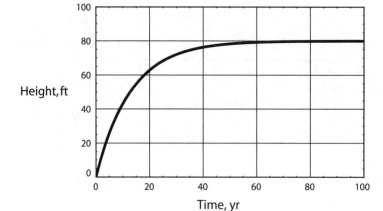

Learn to talk about your graphs and convey their basic messages. Here's how I would talk about mine. This is a picture of rapid initial growth during the first 20 years, eventually slowing as the tree approaches a mature height of 80 feet. The height of the tree in this graph is always increasing until the tree reaches its mature height. However, the height is increasing at an ever-decreasing rate. (We'll get serious about rates of change in the next chapter.) ◇

FOCUSED PROBLEMS :: CHAPTER 1

1. It's 80° F in your kitchen. You preheat your oven to 400° F. Then you put a casserole in the oven. Sketch a graph of the temperature at the center of the casserole as a function of time.

2. Sketch a graph of the height of the milk in your glass as you steadily sip it through a straw.

3. You are filling a glass from the faucet. Water flows into the glass at a steady rate. Sketch a graph of the height of the water in the glass as a function of time. The glass you are thinking of probably looks like the first one shown in Figure 3. Make a graph for that glass, then another for each of the other three glasses shown.

FIGURE 3.
Focused Problem 3.
Glasses of different shapes.

4. You toss a ball up into the air. Sketch a graph of the height of the ball as a function of time, starting from a moment in time just *after* it leaves your hand and ending just *before* it lands back in your hand.

SOLUTIONS CAN BE FOUND ON PAGE 347

CHAPTER 2

Rates of Change

Change is at the heart of physics. That's because physics is the science that seeks to explain why anything happens at all! The physicist wants to know what makes any given situation develop as time progresses. In order to begin to answer this question, it is important to be able to understand change, and rates of change, intuitively as well as quantitatively.

An example of a quantity that changes over time is your own height over the course of your life. Figure 4 shows a rough graph

This may be the most intellectually demanding chapter in the entire book. You don't have to become expert in this material now, but you do have to get the basics right away. And you certainly will have to become expert before long if you want to master physics. Expect to revisit this chapter several times.

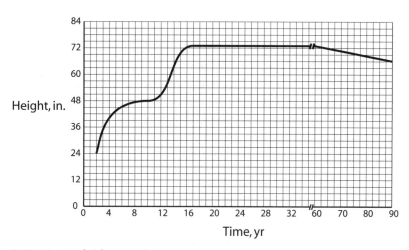

FIGURE 4. My height versus time.

of how my height has changed over time and how it should continue to change over the decades to come (we'll make the hopeful assumption that I live as long as the graph indicates!). Notice the way the horizontal scale of the graph changes at age 36. To the left of this point, each interval represents one year; to the right, each interval represents 2.5 years.

Rate of Growth

Examine the curve at the left-hand side of Figure 4, in the age range from 2 years to 4 years. Looking at the curve there, how would you describe the *rate of growth* between ages 2 and 4?

- You might use a word like "fast." (This kid is shooting up like a weed!)

How would you describe the rate of growth between the ages of 20 and 25?

- You might say, "No growth" or "Zero growth."

How would you describe the rate of growth between the ages of 70 and 90?

- You might say, "Shrinking"—or, even better, "Negative growth." You might also note that the negative growth during this period is "gradual," or "slow," only a few inches over the course of 30 years.

These are all good intuitions, and we want to build on them so that we can be more quantitative.

- Here are the basic terms we use to describe rates of change. When the quantity is going up, we say that the quantity is *increasing*. When the quantity is going down, we say that the quantity is *decreasing*. When the quantity is staying the same, we say that the quantity is, uh, *staying the same*.

» WORKED EXAMPLE 1

Figure 5 shows a graph of soil temperature versus time over a 24-hour period, with $t = 0$ representing the time when the measurements first began to be recorded. During which times is the temperature increasing? During which times is the temperature decreasing? During which times is the temperature staying the same?

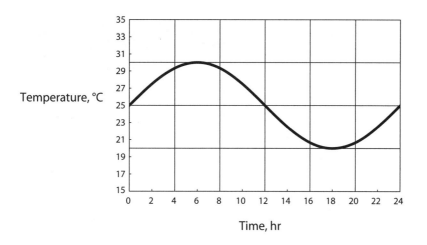

FIGURE 5.
Worked Example 1.
Soil temperature
versus time.

» SOLUTION 1

The temperature is increasing from the beginning of the graph
($t = 0$) until just before $t = 6$ hr. The temperature is also increasing
from just after $t = 18$ hr through the end of the graph ($t = 24$ hr).
The temperature is decreasing from just after $t = 6$ hr until just
before $t = 18$ hr. The temperature is staying the same only momentarily at $t = 6$ hr and $t = 18$ hr. ◇

Rate of Change = Slope of the Graph

There is a simple but profound relationship between the rate at
which a quantity is changing and the appearance of its graph.

- In places where the curve is steep uphill, the quantity is
  increasing rapidly.

- In places where the curve is flat—horizontal—the quantity
  is not changing at all.

- In places where the curve is steep downhill, the quantity is
  decreasing rapidly. In these places, the rate of change is a
  negative number with a large absolute value.

To illustrate the idea that the rate of change equals the slope, let
me show you a rough sketch of the *rate of change* of my height based
on Figure 4. The rate of change of my height is graphed in Figure 6.
I made Figure 6 by simply "eyeballing" Figure 4. Where the height
curve was steep uphill, I made the rate of change in my height large
and positive. Where the height curve was flat or constant, I made
my rate of change zero. Where the height curve was downhill, I
made my rate of change negative.

The rate at which
a quantity is
changing at any
given moment
equals the slope
of the curve at
that moment.

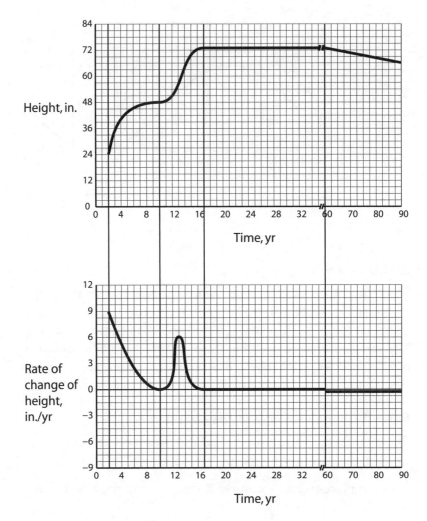

FIGURE 6.
My height versus time (top graph) and the rate of change of my height versus time (bottom graph).

When you get good at this, you can draw accurate graphs of the rate of change just by eyeballing the slope at each point. I like to picture a little hiker trekking across the graph. Where is it difficult for him to climb? There the rate of change is large and positive. Where does he have a steep downhill route? There the rate of change is large and negative. See Figure 7.

Notice that the slope in Figure 7 varies from point to point on the curve. Initially, the slope is very steep. As time goes by, the curve *gradually* flattens out, *momentarily* reaches a flat point, and then *gradually* becomes steeper downhill. In more numerical language, the rate of change starts out large and positive, gradually decreases in value, momentarily equals zero, and then becomes larger and

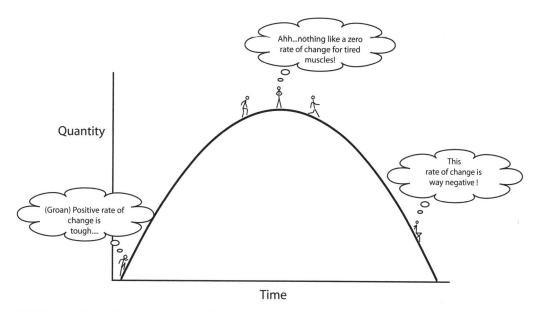

FIGURE 7. A hiker trekking across a graph. The hiker can tell you the slope of the terrain at different points.

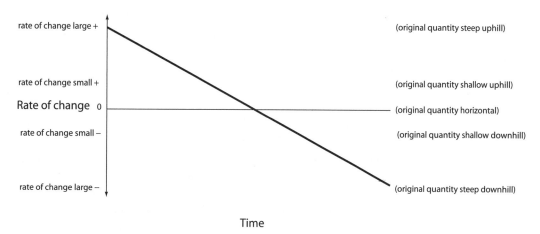

FIGURE 8. The rate of change of the quantity in Figure 7 as a function of time.

larger in the negative range. So a graph of the rate of change of Figure 7 would look like Figure 8.

People are often surprised to see a straight line here. But just follow this diagonal line slowly downward with your eye and reflect on how it faithfully describes the varying slope of the hill in Figure 7, starting positive, then gradually dropping through zero to become negative.

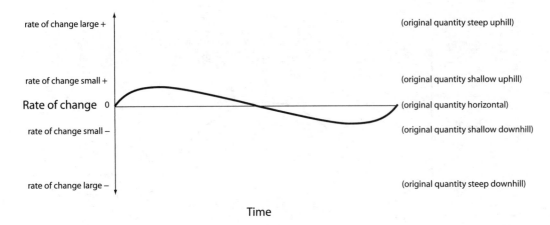

rate of change large +   (original quantity steep uphill)

rate of change small +   (original quantity shallow uphill)

Rate of change   0   (original quantity horizontal)

rate of change small −   (original quantity shallow downhill)

rate of change large −   (original quantity steep downhill)

Time

FIGURE 9.
An incorrect
graph of the
rate of change
of the quantity
in Figure 7.

When I ask students to draw the slope of the hill in Figure 7, they often draw a picture like that in Figure 9. Students tell me they draw this picture because they want the slope to start off at zero. What I do in this case is ask them to look at the first little hiker in Figure 7. Is he standing on flat ground? Not at all! He is standing on very steep ground right at the start. So, when graphing the slope, we should indicate a very steep slope right at the start. There is no rule that says the slope has to start off at zero.

I've noticed that students instinctively place their pencil at zero when they start to make a rate-of-change graph for a quantity like the one shown in Figure 7, perhaps because they see that the quantity itself starts with a value of zero. But don't look at the quantity's *value*—look at its *slope*. When you're new at making rate-of-change graphs, it may help to keep your hands in your lap while you study the curve, making your first mark only after you've decided whether the curve is steep or shallow at the outset.

» WORKED EXAMPLE 2

Figure 10 shows the graph of a quantity versus time. Sketch a rough graph of the rate of change of this quantity versus time by eyeballing the slope at each moment of time.

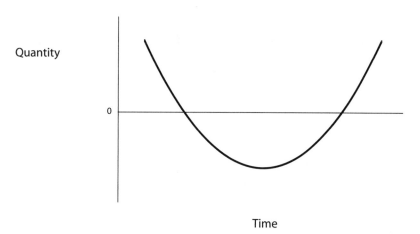

FIGURE 10. Worked Example 2. A quantity versus time.

»SOLUTION 2

Start by looking at the left edge of the curve. At this point, the quantity is *decreasing*. So the rate of change starts off with a *negative value*. Figure 11 shows what we have decided so far.

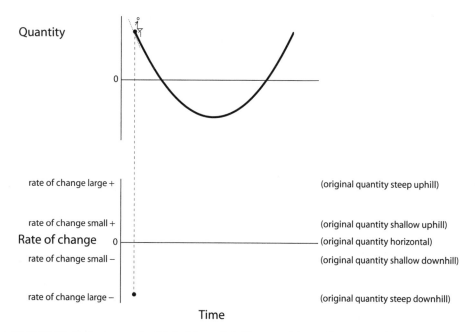

FIGURE 11. Beginning to solve Worked Example 2. The rate of change (bottom graph) of the quantity in the top graph, which is initially negative (steep downhill).

Now move your eye a little to the right on the given curve. The quantity is still decreasing, but the slope is not so steep. Therefore, the rate of change is still negative, but not *so* negative. Figure 12 shows this next step.

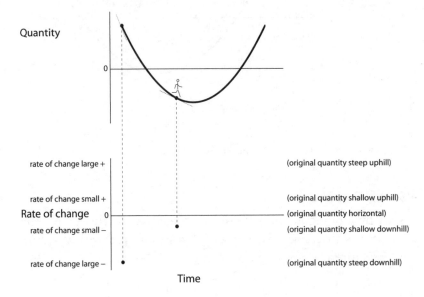

FIGURE 12.
The rate of change (bottom graph) a bit later on, which is a smaller negative number.

Now look at the very lowest point on the graph. Here the graph is (momentarily) horizontal. So the rate of change is (momentarily) zero. Figure 13 shows this conclusion.

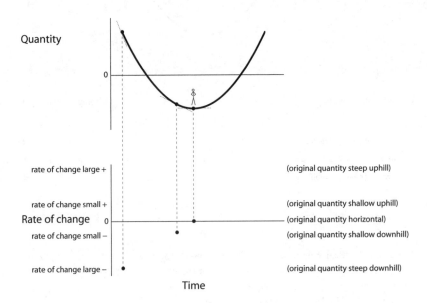

FIGURE 13.
The rate of change (bottom graph) at the bottom of the curve, which is momentarily zero.

Now look at the last part of the graph. Here the quantity is increasing very rapidly. So the rate of change is a large positive number. Figure 14 shows this conclusion.

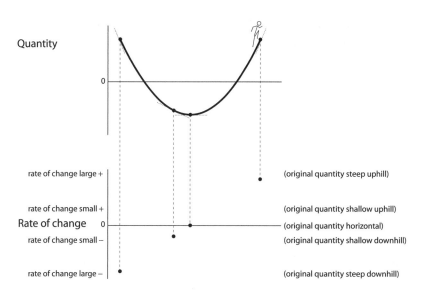

FIGURE 14.
The rate of change (bottom graph) at the end, which is a large positive number.

Now we just connect the dots. Figure 15 shows our best sketch of the rate of change of the quantity.

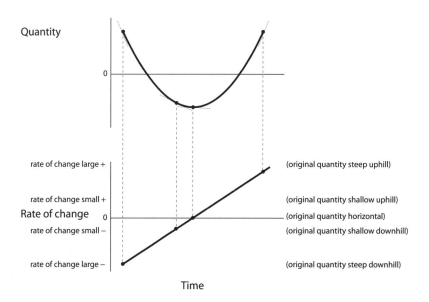

FIGURE 15.
Solution to Worked Example 2. The completed graph of the rate of change (bottom graph) of the quantity in the top graph.

It's really NOT OK if you don't understand this last example. You will be lost for weeks to come unless you can reason in this way. Talk to your instructor if you need help! A little extra investment now will pay big dividends later.

How did I know to draw a perfectly straight line? Really, I didn't. A numerically accurate graph of the rate of change of this quantity *might* have a little curvature to it. To know for sure, we'd have to be given a symbolic formula for the quantity as a function of time, and then we could use calculus to find a symbolic formula for the rate of change as a function of time. However, at the eyeball level of accuracy, a straight line *does* faithfully represent the rate of change of a quantity that depends on time in the way shown in Figure 10. ◇

Those last two rate-of-change graphs both turned out to be straight lines. In the next section we'll see some curved examples, more like that in Figure 6.

Computing Rates of Change Numerically

It's great to be able to eyeball rates of change qualitatively—and you should continue to work on that—but it's also important to be able to calculate rates of change numerically.

The slope is the rise over the run of the tangent line.

Because the rate of change is the slope, what we need is a way to calculate the slope of a curve at a given point in time. The basic rule is simple: The slope of a curve, at any given point, is equal to the rise over the run of the tangent line at the given point.

The best way to explain this rule is by example.

» WORKED EXAMPLE 3

It's OK to eyeball the tangent; we're just looking for a good estimate. One learns how to find the exact tangent line in calculus class; if you're interested, Focused Problem 8 at the end of this chapter shows the basic idea.

Figure 16 shows the price of a company's stock as a function of time over a 48-week period. Estimate the rate at which the price is changing at time $t = 12$ wk.

» SOLUTION 3

Step 1. Look at the graph, and focus on the curve at time $t = 12$ wk. This point is represented in Figure 17 by a large dot.

Step 2. Use a ruler or the straight edge of a piece of paper to draw a line that goes through the dot.

Of course, *many* lines go through the dot! So what you should do is eyeball the line that seems to just barely "kiss" the curve at this single point. You want the line to represent the slope of the curve right in the neighborhood of the dot. This line is called the *tangent line*.

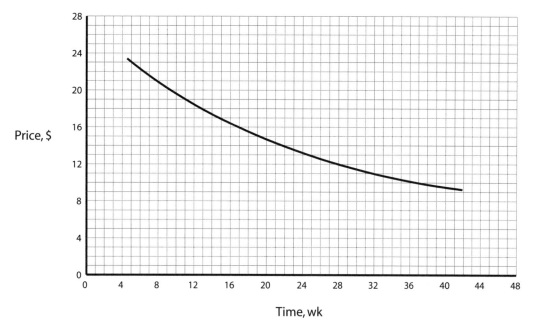

FIGURE 16. Worked Example 3. The price of a company's stock as a function of time.

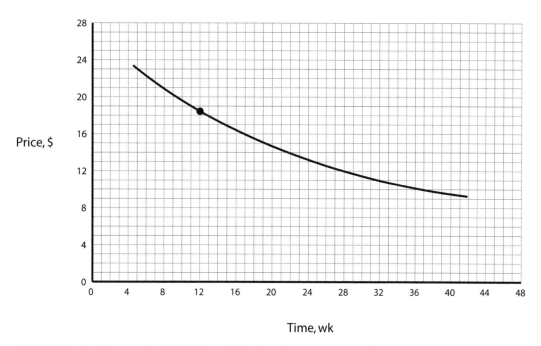

FIGURE 17. Beginning to solve Worked Example 3 by focusing your eye on a single spot. The large dot is where we want to estimate the rate at which the price is changing.

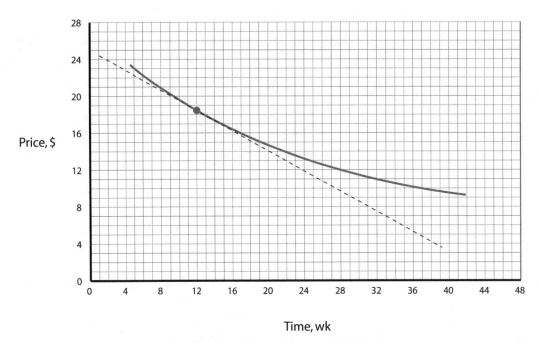

Price, $

Time, wk

FIGURE 18. Adding an estimated tangent line at $t = 12$ wk.

Figure 18 shows my estimate for the tangent line at time $t = 12$ wk.

Step 3. Choose two points on the tangent line that are pretty far apart, and use them to find the ratio of rise over run. The two points I chose are shown in Figure 19. These were convenient choices because they are points where the tangent line passes through grid lines on the graph paper. (Any two chosen points on the tangent line will work, though.)

Looking at the graph, our first chosen point has coordinates (22 wk, 13 dollars). Our second chosen point has coordinates (31 wk, 8 dollars). So the slope of the tangent line is

$$
\begin{aligned}
\text{slope} &= \frac{\text{rise}}{\text{run}} \\[1em]
&= \frac{8 \text{ dollars} - 13 \text{ dollars}}{31 \text{ wk} - 22 \text{ wk}} \\[1em]
&= \frac{-5 \text{ dollars}}{9 \text{ wk}} \\[1em]
&= -0.56 \frac{\text{dollars}}{\text{wk}}.
\end{aligned}
$$

(1)

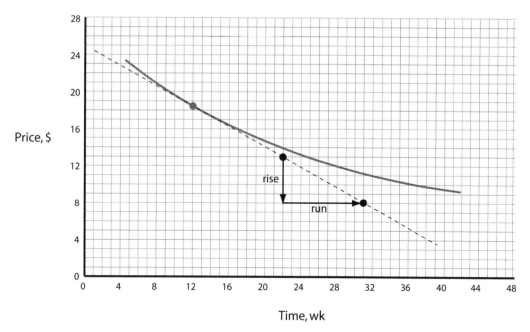

FIGURE 19. Choosing two points from which to calculate the slope of the tangent line.

This means that at time $t = 12$ wk, the stock price is falling at a rate of 56 cents every week. $\diamond$

By calculating numerical rates of change this way, we can put numerical values on our rate-of-change graphs. For example, in Figure 20 I have sketched a graph of the rate of change of the stock price, and I have labeled the graph with the value we just calculated.

The Rate-of-Change Symbol

We'll be talking about rates of change quite a lot, so for the sake of economy it would be nice if we could condense the phrase "the rate of change of" into a single symbol. We'll borrow a symbol from calculus for this: $\frac{d}{dt}$. The symbol $\frac{d}{dt}$ should be read as "rate of change of." Thus, if my height is denoted by H, then the rate of change of my height is denoted by $\frac{d}{dt}H$. Read $\frac{d}{dt}H$ aloud as "the rate of change of height."

Estimating Rates of Change Using Two-Point Data

Earlier we saw how to calculate rates of change numerically by sketching the tangent line to a given curve. But even if you aren't given a curve that shows pictorially how the quantity Q varies over time, you can still estimate the rate of change of Q at a single point

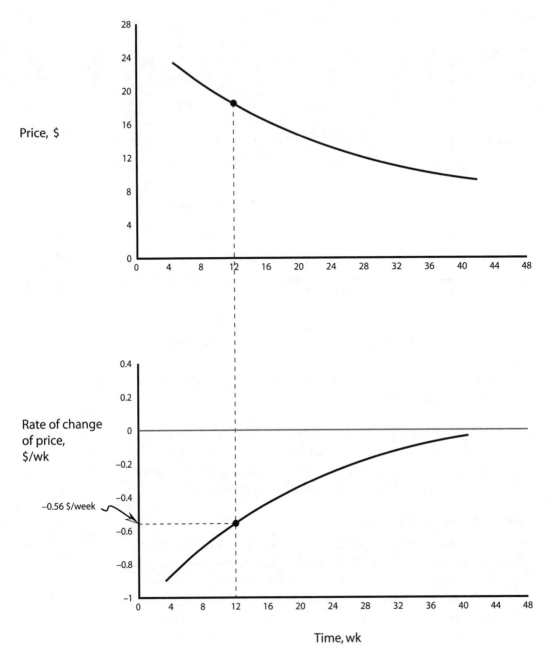

FIGURE 20. Solution to Worked Example 3. The rate of change of the stock price, labeled with numerical values (bottom).

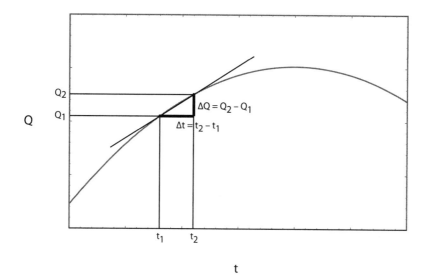

FIGURE 21.
Estimating $\frac{d}{dt}Q$
between t_1 and t_2
using two-point data
by calculating the
ratio $\frac{\Delta Q}{\Delta t}$.

in time as long as you have *two-point data:* in other words, as long as you know two values of Q at two nearby times. If the value of Q at time t_1 is Q_1 and the value of Q at time t_2 is Q_2, then the rate of change of Q between t_1 and t_2 is given approximately by

$$\frac{d}{dt}Q \approx \frac{\Delta Q}{\Delta t} = \frac{Q_2 - Q_1}{t_2 - t_1}. \tag{2}$$

As Figure 21 illustrates, the ratio $\frac{\Delta Q}{\Delta t}$ gives a good approximation to $\frac{d}{dt}Q$ as long as the times t_1 and t_2 are reasonably close together.

» WORKED EXAMPLE 4

At 3:00 p.m., the speed of an ocean liner was 18 knots. At 3:15 p.m., the speed of the ocean liner was 16 knots. Estimate the rate of change of the ocean liner's speed during this time period.

» SOLUTION 4

The problem gives us two-point data as follows:

$$Q_1 = 18 \text{ knots} \tag{3}$$
$$t_1 = 3\!:\!00 \text{ p.m.} \tag{4}$$
$$Q_2 = 16 \text{ knots} \tag{5}$$
$$t_2 = 3\!:\!15 \text{ p.m.} \tag{6}$$

Applying Equation 2, we have

$$\frac{d}{dt}Q \approx \frac{\Delta Q}{\Delta t} = \frac{Q_2 - Q_1}{t_2 - t_1} \tag{7}$$

$$= \frac{16 \text{ knots} - 18 \text{ knots}}{3:15 \text{ p.m.} - 3:00 \text{ p.m.}} \tag{8}$$

$$= \frac{-2 \text{ knots}}{0.25 \text{ hr}} \tag{9}$$

$$= \frac{-2}{0.25} \frac{\text{knots}}{\text{hr}} \tag{10}$$

$$= -8 \frac{\text{knots}}{\text{hr}}. \tag{11}$$

During the period from 3:00 to 3:15 p.m., the ocean liner's speed was decreasing at a rate of 8 knots per hour. ◇

Net Changes

Often we need to find the *net change* in a quantity that varies with time. If you start a month-long exercise program, your weight goes up and down over the course of the month. At the end of the month, you want to know the net change in your weight from start to finish. More generally, if some quantity Q varies with time, then you often need to find the net change in Q over a certain time interval, from some initial time t_i to some final time t_f.

Obviously, if you are looking at a graph that shows the values of Q over time, then you can easily find the net change. Just read off the value of Q at each of the two times, and subtract. See Figure 22.

The fascinating thing is that you can also find the net change in Q by looking at a graph of the *rate of change* of Q! To do that, you compute the area under the rate-of-change curve. Let me state this more carefully, and then I'll explain further:

$$Q_f - Q_i = \text{net signed area bounded by the rate-of-change curve} \tag{12}$$
$$\text{and the } t\text{-axis, from } t_i \text{ to } t_f.$$

The term *signed area* means that we count the area as positive where the rate-of-change curve is above the t-axis, negative where the rate-of-change curve is below the t-axis.

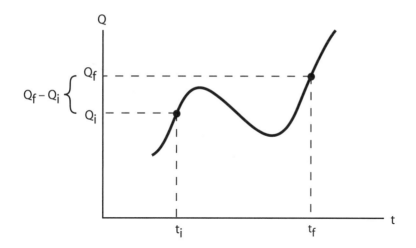

FIGURE 22.
Finding the net
change $Q_f - Q_i$ from
a graph of Q.

Equation 12 will reappear in several different forms later in this book (see Chapter 9, Equations 84 and 89). I will not prove Equation 12 mathematically here—this is done in a first-semester calculus class—but I will offer an intuitive justification, because the intuitions are ones you'll need to have, too.

Equation 12 is known in mathematics as the Fundamental Theorem of Calculus, where it is expressed in symbols as $\int_{t_i}^{t_f} Q'(t)\, dt = Q(t_f) - Q(t_i)$. This theorem was glimpsed by Descartes in 1639, fully articulated independently by Newton and Leibniz in the late seventeenth century, and finally proved to contemporary standards of rigor by Cauchy and others in the nineteenth century [2].

To understand Equation 12 on an intuitive level, it helps to start with a special case. Figure 23 shows the graph of a quantity Q that varies with time at a constant rate, as well as a graph of the rate of change of Q.

In this case, we can reason as follows:

- Because the $Q - t$ graph (top) is a straight line, the slope of the $Q - t$ graph everywhere in the entire interval from t_i to t_f is simply given by the same value of rise/run, or $(Q_f - Q_i)/(t_f - t_i)$. This ratio is the value of the rate of change for the entire interval, as shown in the rate-of-change graph (bottom).

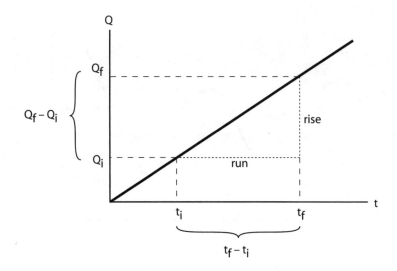

FIGURE 23.
A quantity Q that changes at a constant rate (top) and the rate of change of Q (bottom).

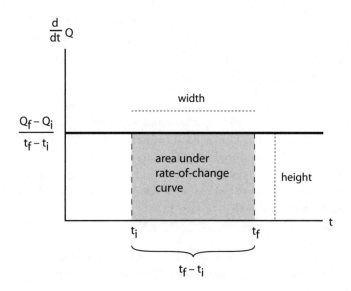

- The shaded area under the rate-of-change curve is rectangular in shape. So its area is given simply by (height times width). This is

(13)   area under rate-of-change curve = height × width

(14) $$= \left(\frac{Q_f - Q_i}{t_f - t_i} \right) \times (t_f - t_i)$$

(15)   area under rate-of-change curve $= Q_f - Q_i$.

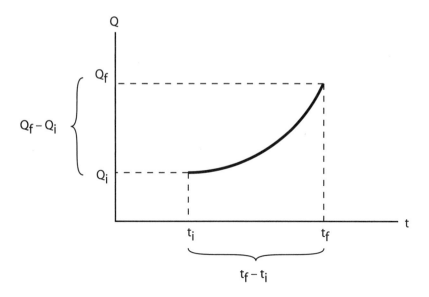

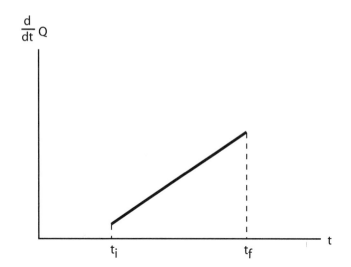

FIGURE 24.
A quantity Q that changes at a variable rate (top) and the rate of change of Q (bottom).

So in this case—when the rate of change is constant—it is easy to see that the net change in Q is equal to the area under the rate-of-change curve. Cool, huh?

What if the rate of change of Q is not constant? That is, what if Q changes at a varying rate, as in Figure 24?

To handle this case, we can approximate the $Q - t$ curve as a sequence of straight segments. See Figure 25.

Multiplying to find the area simply "undoes" the division we did in the first place to find the slope.

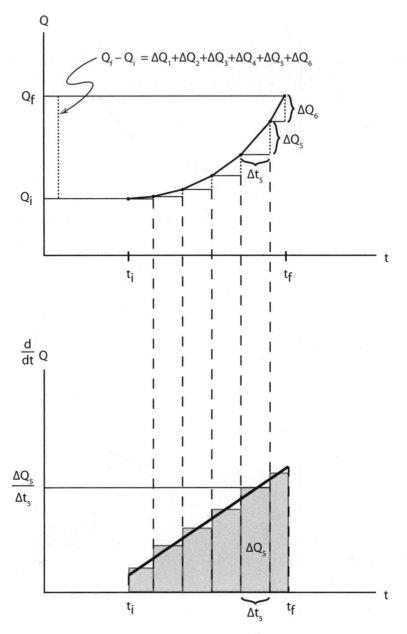

FIGURE 25. Approximating the original $Q - t$ curve in Figure 24 (top graph) by means of a sequence of straight segments (segments from the top graph have rates of change graphed at the bottom).

Let the number of segments be as large as you like, and the width of the
segments be as narrow as you like, until you are comfortable with the
accuracy of the argument that comes next.

To find the net change $Q_f - Q_i$ from t_i to t_f, we can add up all of
the small changes ΔQ_1, ΔQ_2, ΔQ_3, ΔQ_4, ΔQ_5, and ΔQ_6 along the
way. (See the top graph in Figure 25. ΔQ_5 and ΔQ_6 are labeled at
right. The other ΔQ_ns are not labeled in order to keep the diagram
free of clutter.) Now, by our earlier reasoning in Figure 23, each
small change ΔQ_n will be equal to the area of its corresponding
shaded rectangle in the rate-of-change graph below. So adding up
all of the small changes ΔQ_n means adding up all of the rectangular
areas. But adding up all of the rectangular areas just means finding
the total area under the rate-of-change curve. So again, the net
change from t_i to t_f is equal to the area under the rate-of-change
curve. This completes our intuitive analysis of Equation 12. The
conclusion bears repeating:

$$Q_f - Q_i = \text{net signed area bounded by the rate-of-change curve}$$
$$\text{and the } t\text{-axis, from } t_i \text{ to } t_f. \tag{16}$$

Textbooks usually provide an end-of-chapter summary for you. While
that may *seem* helpful, what it actually does is deny you the chance
to synthesize the material on your own—a crucial part of the learning
process. Sadly, there are no shortcuts in learning physics. On the next
page you can solidify what you've read by making your *own* summary.

Reader's Notes

Feel free to fill the space below with words, equations, and diagrams. And remember, digesting the chapter is just a start; only by doing the Focused Problems can you be sure that you know how to apply the concepts yourself.

FOCUSED PROBLEMS :: CHAPTER 2

1. Which number is bigger, −8 or −9? Can you argue the question both ways?

2. Sketch graphs that illustrate the following kinds of quantities and their behaviors. (If you like, you can simply sketch one long curve that exhibits all of the behaviors in sequence.)

   a. A quantity that increases *at a steady rate* over time.

   b. A quantity that increases *at an increasing rate* over time.

   c. A quantity that increases *at a decreasing rate* over time.

   d. A quantity that *remains constant* over time.

   e. A quantity that decreases *faster and faster* over time.

   f. A quantity that decreases *slower and slower* over time.

   g. A quantity that decreases *at a steady rate* over time.

3. For each of the following situations (familiar from Chapter 1), sketch a non-numerical graph of the given quantity as a function of time. Then, immediately below that, make a graph of the *rate of change* of the quantity as a function of time. Align the graphs carefully.

   a. It's 80° F in your kitchen. You preheat your oven to 400° F. Then you put a casserole in the oven. Sketch a graph of the temperature at the center of the casserole as a function of time.

   b. Sketch a graph of the height of the milk in your glass as you steadily sip it through a straw.

   c. You are filling a glass from the faucet. Water flows into the glass at a steady rate. Sketch graphs of the height of the water in each of the four glasses shown in Figure 26 as a function of time.

FIGURE 26. Focused Problem 3(c). Glasses of different shapes.

   d. You toss a ball up into the air. Sketch a graph of the height of the ball as a function of time, starting from a moment in time just *after* it leaves your hand and ending just *before* it lands back in your hand.

4. For each of the situations you considered in the last problem, sketch a non-numerical graph of the *rate of change of the rate of change* of the quantity as a function of time. (I would do it like this: Graph the quantity, then graph the rate of change of that, and then graph the rate of change of that, aligning the three graphs vertically.)

5. Going back to Worked Example 3 in this chapter, use Figure 16 to numerically estimate the rate at which the stock price is changing at time $t = 30$ wk.

6. According to Jürgen Giesen's calculations [3], on September 20, 2006, there were 12.12 hours of daylight in Bennington, Vermont. On September 27, 2006, there were 11.78 hours of daylight. Estimate the rate at which the length of the day was changing during this time period. Express your answer in hours per day and also in minutes per day.

7. Figure 27 shows the rate of change of the temperature of the water in a puddle over an 8-hour period. What was the net change in temperature over the following time periods?

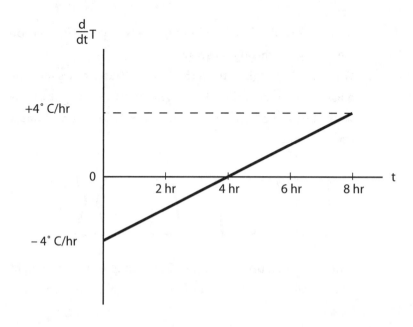

FIGURE 27.
Focused Problem 7. The rate of change of the temperature of the water in a puddle over an 8-hour period.

a. From $t = 0$ to $t = 2$ hr

b. From $t = 0$ to $t = 4$ hr

c. From $t = 0$ to $t = 6$ hr

d. From $t = 0$ to $t = 8$ hr

e. If the temperature of the puddle was $30°$ C at $t = 0$, what was the temperature of the water at its coldest point?

f. Sketch a graph of the temperature versus the time from $t = 0$ to $t = 8$ hr. Label the vertical axis with numerical temperature values based on your answers from previous questions.

8. A quantity Q changes over time t according to the formula $Q(t) = 10 + 4t - t^2$, where t is in seconds. See Figure 28.

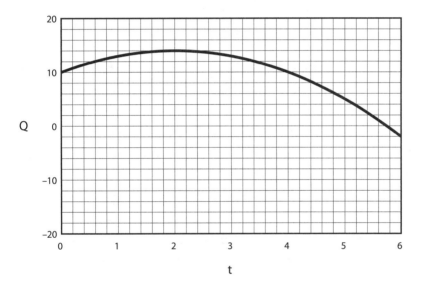

FIGURE 28.
Focused Problem 8.
The quantity
$Q(t) = 10 + 4t - t^2$
from $t = 0$ to $t = 6$.

Q

t

a. Use the graph to estimate $\frac{d}{dt}Q$ at time $t = 3$.

The *exact* value of $\frac{d}{dt}Q$ at time $t = 3$ can be found in the following way. First, identify two particular points on the curve, a point P_1 at the time $t = 3$ we are focusing on and a point P_2 nearby. See Figure 29. The coordinates of these points are $P_1 = (3, Q(3))$ and $P_2 = (3 + \Delta t, Q(3 + \Delta t))$. Here Δt represents a small increment of time.

b. Use the given formula, $Q = 10 + 4t - t^2$, together with the coordinates of P_1 and P_2, to show that the slope of the line connecting P_1 and P_2 is given by $\frac{\Delta Q}{\Delta t} = -2 - \Delta t$.

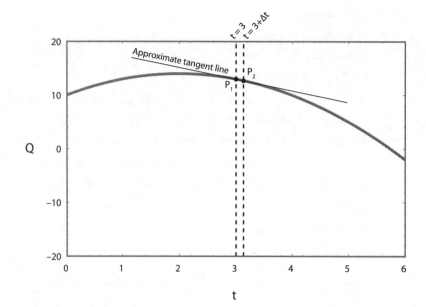

FIGURE 29.
Beginning to solve Focused Problem 8. The thin solid line through points P_1 and P_2 is a good estimate of the tangent line at time $t = 3$, when the increment Δt is small.

c. The smaller Δt is, the better the tangent line in Figure 29 "kisses" the curve. So if we put $\Delta t = 0$, we'll have the exact tangent line! Setting Δt to zero in the answer from part (b), we find the exact rate of change at time $t = 3$ to be $\frac{d}{dt}Q = -2$. Is this value close to the rate of change you estimated in part (a)?

Next we'll consider the rate of change at any chosen time t instead of just at the specific time $t = 3$.

d. Show that at any chosen time t, the rate of change of Q at time t is given by $4 - 2t$. (Hint: Adapt your method from parts (b) and (c).)

e. Graph the rate-of-change function $\frac{d}{dt}Q = 4 - 2t$ from $t = 0$ to $t = 6$. Is the graph a plausible description of the slope of the curve in Figure 28?

SOLUTIONS CAN BE FOUND ON PAGE 349

CHAPTER 3

Introducing Position and Velocity

The three basic concepts in the study of motion are position, velocity, and acceleration. We begin our study of motion in earnest with the first two of these concepts: position and velocity.

Position

The concept of position is easy to summarize. Position is where something is at any given time. But there is more than meets the eye in this simple concept. Consider what we say when we talk about the position of something in everyday speech:

> "The boat is 25 miles due west of the lighthouse."

> "The bird is 1,000 feet above sea level."

> "The treasure is 120 miles from here, at a bearing of 30° north of east."

These are the kinds of sentences we use to specify the positions of things. Notice that these sentences have a few characteristics in common. What are they?

All of the sentences specify the following three things:

- How far
- In what direction
- From where or from what point.

This chapter is a first pass at position and velocity; we'll return to these concepts again right after we've studied vectors in Chapter 4.

Look for these elements in the original sentences.

"The boat is <u>25 miles</u> <u>due west</u> of <u>the lighthouse</u>."

"The bird is <u>1,000</u> feet <u>above</u> <u>sea level</u>."

"The treasure is <u>120 miles</u> from <u>here</u>, at a bearing <u>30° north of east</u>."

The Origin

To talk about the position of something, you need some point of reference (such as the lighthouse, sea level, or "here"). This point of reference is called the *origin*.

Representing Position with a Diagram

Let's think a little bit more about the sentence that told us the position of the treasure. We can draw a picture to represent the position. See Figure 30.

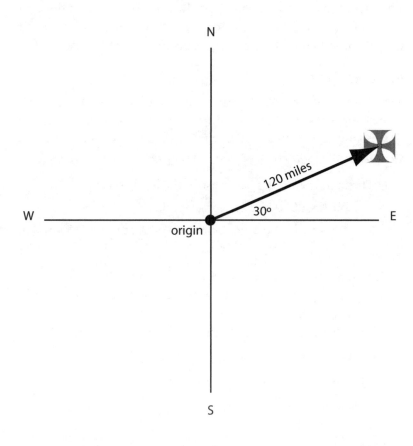

FIGURE 30.
The position of a treasure described as being "120 miles from here, at a bearing 30° north of east."

I have drawn an arrow on the treasure map. An arrow is a good way to represent the position of something. The length of the arrow tells us how far away the object is from the origin, and the orientation (or angle) of the arrow tells us the direction.

We could call this arrow "the position arrow," but technically these arrows are known as vectors. The arrow drawn on the treasure map is called *the position vector* (of the treasure).

The arrow convention for position can sometimes be misleading. To some students, an arrow intuitively suggests a *journey*, one that begins at the origin and ends at the location of the object. But no "journey" is being indicated in Figure 30. The position vector does not indicate where an object has been in the past or where it will be in the future; it indicates only the place where the object stands at the present moment.

We'll be seeing vectors all the time in our study of physics. Position, velocity, acceleration, force, and momentum are all well described by vectors. By their nature, all of these things have a "size" as well as a direction associated with them. (We'll look at this more in Chapter 4.)

Because vectors are basically arrows, we often use arrow symbols when we write vector quantities to emphasize their vectorial status. For example, we might call the treasure's position vector $\vec{r}$. Arrow symbols like this are usually used in handwriting. In typed or printed documents like this book, vectors are usually named using boldface letters. So in this book we'll use a symbol like **r** to represent the position of the treasure.

Whenever you see a boldface letter in this book, like **r**, **v**, **a**, **F**, or **p**, remember that you're looking at an arrow!

Position versus Distance

Position is not the same as our everyday term *distance*. Distance refers only to "how far" away something is from the origin. But the position of an object refers *both* to how far away it is from the origin *and* to its direction from the origin. This is a lot of information. Distance from the origin is just one ingredient in the position of an object. See Figure 31.

To summarize: Position is where something is at any given time. To specify the position of an object, you first need to specify an origin. Then the position of the object can be represented by a vector. The *length* of the vector tells how far away the object is from the origin. The *direction* of the vector tells which way it is from the origin.

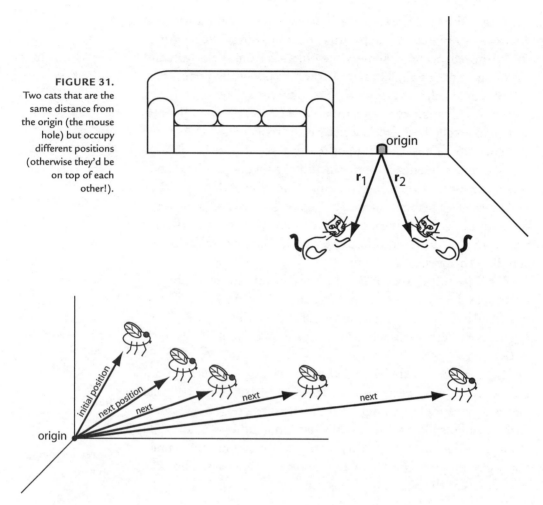

FIGURE 31.
Two cats that are the same distance from the origin (the mouse hole) but occupy different positions (otherwise they'd be on top of each other!).

FIGURE 32.
"Snapshots" of a fly, equally spaced in time. Can you "animate" this series in your mind's eye?

Motion

Motion refers to being at different positions at different times. Figure 32 shows the motion of a buzzing fly. When the position of the fly changes by a lot in a given amount of time, it means the fly is moving fast.

Velocity

Velocity is the second of our three major tools for describing motion. Just as with position, we can learn something about the concept of velocity by examining how we talk about it in everyday speech. Consider the following sentences:

"The suspect is headed east at approximately 95 miles per hour."

"The diver is heading downward at 10 meters per second."

What do these two statements have in common? I notice that both of the statements specify two things:

- How fast?
- In what direction?

Here are those elements emphasized:

"The suspect is headed <u>east</u> at approximately <u>95 mph</u>."

"The diver is traveling <u>downward</u> at <u>10 meters per second</u>."

Velocity describes the way an object is moving at a given instant of time: how fast the object is moving and in what direction.

Velocity versus Speed

Velocity is not the same as our everyday term *speed*. Speed refers only to "how fast" something is going. But the velocity of an object refers *both* to how fast it's going *and* in what direction it's going. This is a lot of information. As you can see from Figure 33, arrows offer us a nice way to represent both aspects of velocity. The length of a velocity vector indicates how fast the object is moving, and the direction of the velocity vector tells the direction of its motion.

The two spiders crawling in the bathtub represented in Figure 33 have the same speed; they are both moving at a speed of

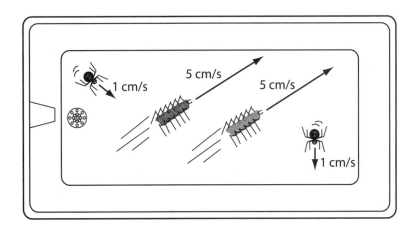

FIGURE 33.
Two spiders and two centipedes crawling in a bathtub. The two spiders have the same speed but are headed in different directions. But the two centipedes have the same speed *and* are headed in the same direction.

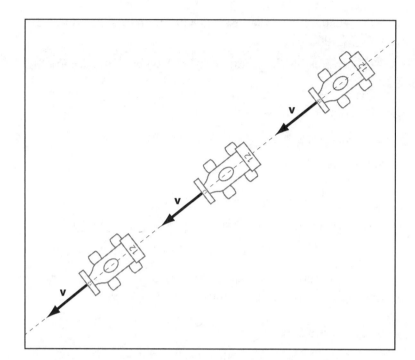

FIGURE 34.
A car whose velocity vector never changes. The "snapshots" of its motion are equally spaced in time.

1 centimeter per second. But their velocity vectors are different because the spiders are headed in different directions. See how the arrows are different? On the other hand, the two centipedes *do* have the same velocity vectors as one another. They have the same speed, *and* they are headed in identical directions. See how the arrows are the same?

Curvilinear Motion

Consider a car with a given velocity vector **v** (Figure 34). If the *direction* of the car's velocity vector never changes, this means that the car follows a straight-line path. If, in addition, the *length* of the car's velocity vector never changes, this means that the car maintains a constant speed.

Of course, most moving objects *do not* follow a straight-line path, and they *do not* maintain a constant speed over time. In general, objects move in curved paths at varying speeds. Just think of a fly buzzing around a bowl of peaches. See Figure 35.

The Position Vector in Curvilinear Motion

Figure 36 shows the position vector **r** of the buzzing fly at three instants of time. Because the fly is moving, the position vector will

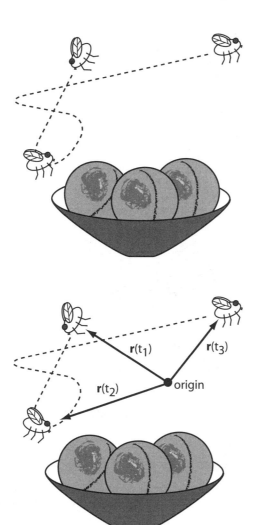

FIGURE 35.
A fly buzzing around a bowl of peaches. Its path is curved, and its speed also varies over time.

FIGURE 36.
The position vector of the fly from Figure 35 at three instants of time.

depend on time, as $\mathbf{r}(t)$. The tip of the position vector follows the fly as it moves through the air. The curve traced out by the fly as it moves through space is called the fly's *trajectory*. In Figures 35 and 36, the fly's trajectory is shown as a dashed curve.

The Velocity Vector in Curvilinear Motion

It is important to be able to accurately sketch the velocity vector of a moving object. To do this well, there are really only three things to keep in mind:

1. Draw the velocity vector "attached" to the moving object.
2. Draw the velocity vector tangent to the trajectory.

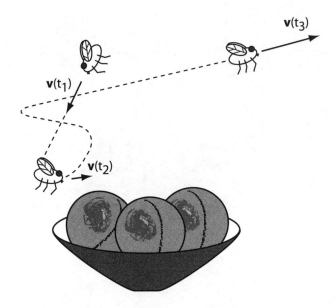

FIGURE 37.
The velocity vector of the fly from Figure 35 at three different instants of time. Try to animate the motion in your mind's eye.

3. Make the length of the velocity vector *roughly* representative of how fast the object is moving. (Just use your intuition for this.)

Figure 37 shows the velocity vector for our fly at three different instants of time.

The Formal Definition of Velocity

Intuitively speaking, an object is moving fast when its position changes rapidly. So it may come as no surprise to learn that velocity is formally defined as the rate of change of position. Let me emphasize this as follows:

(17)   The velocity of an object is the rate of change of its position.

This is written using mathematical symbols as

$$\mathbf{v} = \frac{d}{dt}\mathbf{r}.$$

(18)

Remember from Chapter 2 that $\frac{d}{dt}$ is a single mathematical symbol that means "the rate of change of." So Equation 18 is read aloud as follows: "Velocity is the rate of change of position." We'll be working with this equation again in Chapter 5.

Reader's Notes

Below is space for you to summarize key points from this chapter.

FOCUSED PROBLEMS :: CHAPTER 3

1. Describe a real-life scenario in which an object's position vector varies with time but its distance from the origin does not. Draw some snapshots of the motion to show the position vectors at different times.

2. Describe a real-life scenario in which an object's velocity vector varies with time but its speed does not. Draw some snapshots of the motion to show the velocity vectors at different times.

3. For each snapshot of the fly in Figure 32, sketch the fly's velocity vector at that instant of time. How does the fly's speed vary over the course of the motion?

SOLUTIONS CAN BE FOUND ON PAGE 360

CHAPTER 4

Vectors

In Chapter 3 I explained two kinds of vectors, position (**r**) and velocity (**v**). We'll be working with vectors constantly in this book, so let's take some time now to learn how to analyze them in detail.

The Magnitude and Direction of Vectors

A "vector quantity" in physics is any quantity that, by its nature, has a magnitude of sorts, as well as a direction in which it points. Basically, a vector quantity is something for which an arrow representation makes good sense. Table 1 shows some examples of vector quantities in physics.

Adding Vectors to One Another

Sometimes it makes sense to combine two vectors. For example, two people might be pushing a stalled car, one person pushing with a force $\mathbf{F}_1$ and the other person pushing with a force $\mathbf{F}_2$. See Figure 38. We can find the combined effect on the car by combining the two applied forces, that is, by adding the two vectors $\mathbf{F}_1$ and $\mathbf{F}_2$. We'll see more on the subject of combining forces in Chapter 11.

Arrows are not numbers, so how do we "add" them? The method used by Isaac Newton in his *Principia* is a visual method called the *parallelogram law*. Here is how it works. Suppose you're given two vectors, **u** and **v**, as drawings of arrows, as in Figure 39(a). To

TABLE 1. Some Vector Quantities in Physics

Vector quantity	Symbol for vector	Symbol for *magnitude* of vector	Interpretation of magnitude of vector	Units in which magnitude of vector is measured
Position	**r**	r	r = distance from the origin	meters (m)
Velocity	**v**	v	v = speed of a moving object	meters per second (m/s)
Acceleration	**a**	a	a = magnitude of acceleration	meters per second per second (m/s^2)
Force	**F**	F	F = strength of a push or pull	newtons (N)
Momentum	**p**	p	p = amount of "oomph" possessed by a moving object	kg · m/s

FIGURE 38.
Two people applying force to a car. To find the net effect, we may simply add the two vectors.

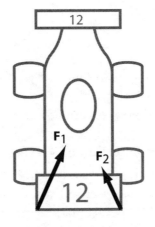

draw the arrow that represents the vector sum of **u** and **v**:

1. Complete the parallelogram defined by **u** and **v** in Figure 39(b).

2. Draw an arrow out to the far corner of the parallelogram in Figure 39(c). This arrow is the vector sum of **u** and **v**.

Position vectors are always based at the origin, but other kinds of vectors may be based at different points. For example, think about the two people pushing on the stalled car in Figure 38. The force vector $\mathbf{F}_1$ is naturally viewed as being "attached" to the place where person 1 is pushing, and force vector $\mathbf{F}_2$ is naturally viewed as being "attached" to the place where person 2 is pushing. If you're trying to use the parallelogram law to add two vectors that are based at different points, you'll have to mentally "slide" one of

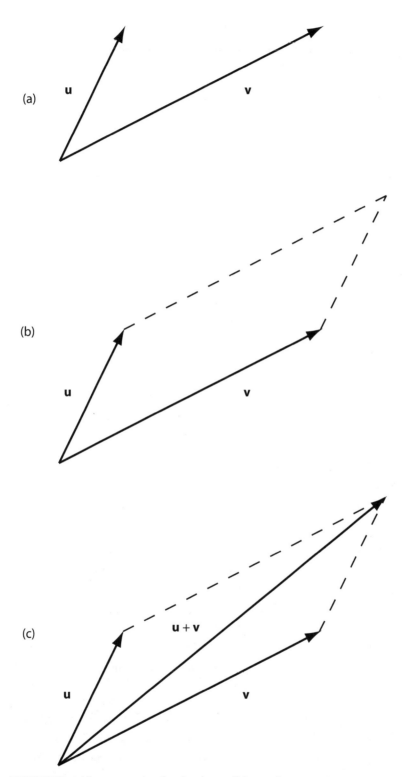

(a)

(b)

(c)

FIGURE 39. Adding vectors visually using the parallelogram law.

the arrows across the page until the base points of the two arrows match up. Keep the direction of the arrow fixed while you slide it.

»WORKED EXAMPLE 1

Given the vectors **u**, **v**, and **w** in Figure 40, use the parallelogram law to draw the vector sums **u** + **v**, **u** + **w**, and **v** + **w**.

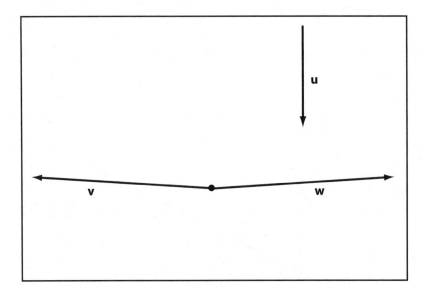

FIGURE 40.
Worked Example 1.
Three vectors.

»SOLUTION 1

The three vector sums are shown in Figure 41. ◇

You probably noticed that the vector sum **v** + **w** is a very small vector—that is, a vector with a very small magnitude. This is because the vectors **v** and **w** are of similar length and point in nearly opposite directions. Under these circumstances, the vectors mostly "cancel out" when you add them.

Another visual method for adding vectors is called the *tip-to-tail method*. In this method, to draw the sum **u** + **v** you slide one arrow until its "tail" lies right at the "tip" of the other arrow. Then you draw an arrow from the first tail to the last tip. See Figure 42.

The tip-to-tail method is especially helpful when the vectors to be added lie in the same straight line, as in Figure 43. In this case it's hard to visualize the parallelogram law, but it's easy to use the tip-to-tail method.

Try adding the vectors in Figure 40 using the tip-to-tail method. You should find that the sums are the same no matter which method you use.

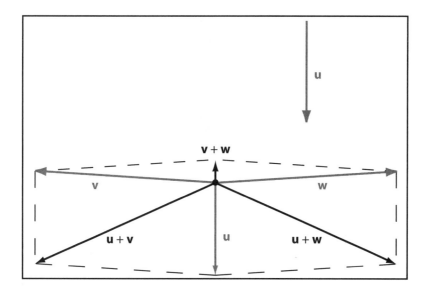

FIGURE 41.
Solution to Worked Example 1. Adding the three vectors from Figure 40 (shown in gray). The three vector sums are shown in black.

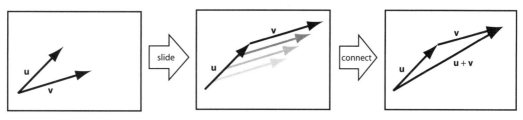

FIGURE 42. Adding vectors visually using the tip-to-tail method.

FIGURE 43.
Worked Example 2. Two vectors in the same straight line.

» WORKED EXAMPLE 2

Given the vectors **u** and **v** shown in Figure 43, use the tip-to-tail method to draw the vector sum **u** + **v**.

» SOLUTION 2

The vector sum **u** + **v** is shown in Figure 44. (For clarity, the vectors are shown offset from one another.) As a first step, I slid the tail

Solution to Worked
Example 2. Adding
vectors **u** and **v**
tip-to-tail.

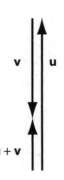

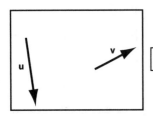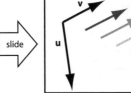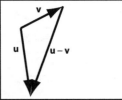

FIGURE 45. Subtracting vector **v** from vector **u** by connecting the tips from **v** to **u**.

of **v** up to the tip of **u**. Then I drew the sum vector **u** + **v** by connecting the tail of **u** to the tip of **v**. This example shows how vectors opposing each other partially cancel when added together.

Subtracting Vectors from One Another

We have seen two ways to draw the vector sum **u** + **v**: the parallelogram law and the tip-to-tail method. It is just as easy to draw the vector difference **u** − **v**. Slide one of the vectors so that the two tails match up, and then draw an arrow from the tip of **v** to the tip of **u** (Figure 45).

» WORKED EXAMPLE 3

Given the vectors **u**, **v**, and **w** shown in Figure 40, draw the vector differences **u** − **v**, **u** − **w**, **v** − **w**, and **w** − **v**.

» SOLUTION 3

The vector differences are shown in Figure 46. ◇

Multiplying Vectors by Ordinary Numbers

Ordinary numbers can be used to "stretch" or "shrink" a vector. For example, if you multiply a vector by the number 3, the vector becomes three times longer. If you multiply a vector by the number 0.5, the vector shrinks to half as long. See Figure 47.

To draw **u** − **v**, connect the tips from **v** to **u**. This ensures that addition and subtraction "undo" each other the way they should. For example, think about what happens in Figure 45 when we add the two vectors **v** and (**u** − **v**) using the tip-to-tail rule. We get **u**, as we should. After all, we want our system of vector math to obey the sensible relation **v** + **u** − **v** = **u**.

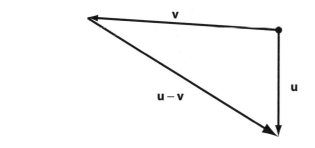

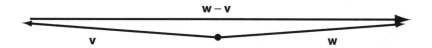

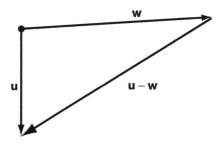

FIGURE 46. Solution to Worked Example 3. The differences of the vectors in Figure 40.

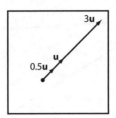

Multiplying a vector by a negative number has the same stretching or shrinking effect, but it also reverses the direction of the vector. For example, if you multiply a vector by the number -0.5, the vector shrinks to half as long and also reverses direction. See Figure 48.

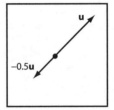

When you multiply a vector by an ordinary number, you are "scaling" the length of the vector (as well as possibly reversing its direction). The scaling action of numbers is why ordinary numbers are technically known as *scalars* when we are speaking in a vectorial context. Many quantities in physics do not have directions associated with them and are therefore well represented by scalars. Table 2 lists some scalar quantities in physics.

By the way, just to warn you about a matter of notation that will be coming up in Chapter 5: Sometimes you'll see a scalar

TABLE 2. Some Scalar Quantities in Physics

Scalar quantity	Symbol	Units
Mass	m	kilograms (kg)
Time interval	Δt	seconds (s)
Energy	E	joules (J)
Electric charge	q	coulombs (C)
Voltage	V	volts (V)

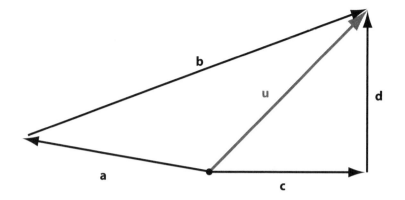

FIGURE 49.
Two ways to realize the vector **u** as a sum of two other vectors: $\mathbf{u} = \mathbf{a} + \mathbf{b}$ and $\mathbf{u} = \mathbf{c} + \mathbf{d}$.

multiplication such as 0.5**u** written as a "scalar division,"

$$\frac{\mathbf{u}}{2}. \tag{19}$$

The two forms 0.5**u** and $\frac{\mathbf{u}}{2}$ are entirely equivalent.

Expressing Vectors as Sums

Let me introduce this topic by making an analogy with ordinary real numbers. There are lots of ways to express the number 40 as a sum of two other numbers. For example, you could write $40 = 20 + 20$, $40 = 10 + 30$, or $40 = -8.1 + 48.1$. In much the same way, any given vector can be expressed in many different ways as a sum of two other vectors. For example, Figure 49 shows two different ways to express, or realize, the vector **u** as a sum of two other vectors. The figure shows the two realizations $\mathbf{u} = \mathbf{a} + \mathbf{b}$ and $\mathbf{u} = \mathbf{c} + \mathbf{d}$. You can use the tip-to-tail method to verify that these sums are correct.

Some realizations of a vector are more useful than others. The most useful realizations are the ones that realize a vector as a sum of two *perpendicular* vectors. The realization $\mathbf{u} = \mathbf{c} + \mathbf{d}$ in Figure 49 is an example of this. By the way, there is more than just one way to realize a given vector as a sum of two perpendicular vectors. See Figure 50.

Coordinate Systems

To solve any sort of quantitative problem in physics, you will have to lay down a *coordinate system*. A coordinate system labels each point in the plane with a pair of real numbers (x, y). (We'll mostly be working in two dimensions.) In Figure 51, point P has x, y

FIGURE 50.
Two ways to realize
the vector **u** as a sum
of two *perpendicular*
vectors: **u** = **c** + **d**
and **u** = **e** + **f**. There
are many more ways
besides.

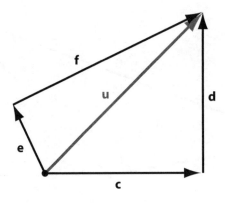

FIGURE 51.
Two different
coordinate systems
for the plane shown
in Figure 50.

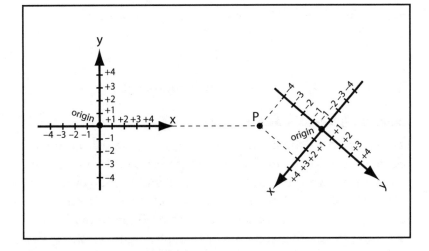

coordinates $(+12.8, 0)$ according to the first coordinate system and x, y coordinates $(+3, -4)$ according to the second coordinate system. One of the coordinate systems is "tilted" relative to the edges of the paper; coordinate systems with a tilt can sometimes be very useful in solving problems.

To lay down a coordinate system of your own, draw a pair of perpendicular number lines, or axes. Values on one axis are labeled x, and values on the other axis are labeled y. The origin is the point where both $x = 0$ and $y = 0$. The arrowheads on each axis indicate the direction of increasing x or y values. When you draw a coordinate system, you have to make sure the two axes are perpendicular to one another. But everything else is up to you. The location of the origin is up to you. The tilt of the axes relative to the edges of the paper is up to you. The arrowheads that show the directions of increasing x and y values are up to you. You'll gradually learn to

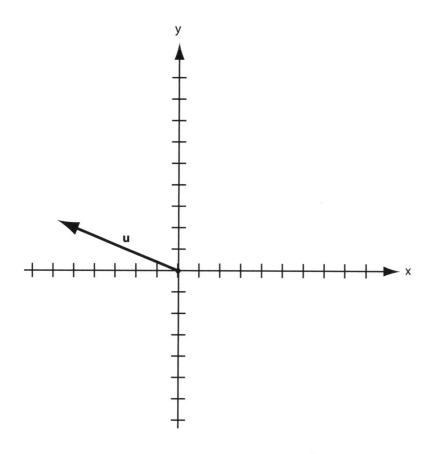

FIGURE 52.
A vector **u** with
its tail at the origin of
a coordinate system.

make these choices in such a way that the problem you're working on becomes easier to solve.

The *x*- and *y*-Components of a Vector

Picture a vector **u** with its tail at the origin of an x, y coordinate system (Figure 52). I like to think of the vector **u** as "casting shadows" onto the two coordinate axes, as shown in Figure 53. The length of the *x*-shadow in Figure 53 is 5 units. The length of the *y*-shadow in Figure 53 is 2 units.

The *x*-shadow in Figure 53 extends for five tick marks in the negative *x*-direction. We express this in words by saying that the "*x*-component" of the vector is −5. The negative sign indicates that the *x*-shadow falls on the negative half of the *x*-axis.

The *x*-component of vector **u** is usually written as u_x. So in this example we would write $u_x = -5$. Similarly, the *y*-component of

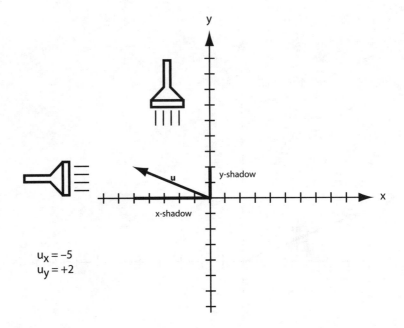

FIGURE 53.
The "shadows cast"
by the vector **u** on the
coordinate axes.

u in this example is $u_y = +2$. The plus sign indicates that the y-shadow falls on the positive half of the y-axis. Let's see another example.

» WORKED EXAMPLE 4

What are the x- and y-components of the vector **v** shown in Figure 54?

» SOLUTION 4

The x-shadow extends for 11 units in the positive x-direction, so we write $v_x = +11$. The y-shadow extends for 5 units in the negative y-direction, so we write $v_y = -5$. ◇

When the coordinate system is tilted relative to the edges of the paper, the procedure is the same. Let's see how that looks.

» WORKED EXAMPLE 5

What are the x- and y-components of the vector **w** shown in Figure 55?

» SOLUTION 5

The x-shadow extends for 11 units in the negative x-direction, so we write $w_x = -11$. The y-shadow extends for 4 units in the negative y-direction, so we write $w_y = -4$. ◇

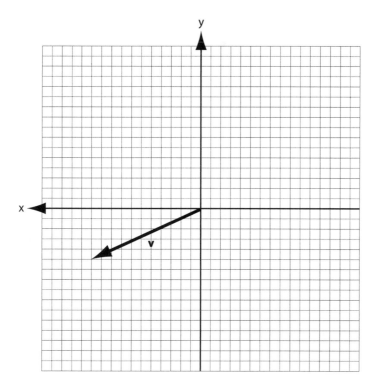

FIGURE 54.
Worked Example 4.
Determining the x-
and y-components of
the vector v.

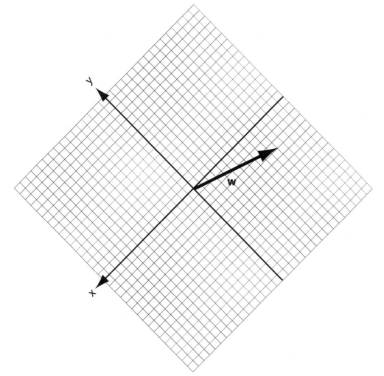

FIGURE 55.
Worked Example 5.
Determining the x-
and y-components of
the vector w.

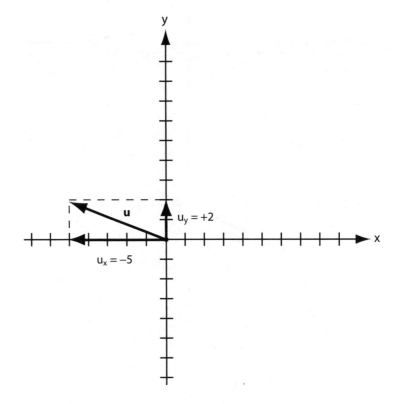

FIGURE 56.
Assembling a vector **u**
from its components
$u_x = -5$ and $u_y = +2$.

Assembling a Vector from Its Components

You can think of the components of a vector as being like "in-structions" that tell you how to assemble the vector (Figure 56). To assemble the vector **u** from its numerical components u_x and u_y, you first draw a vector parallel to the x-axis of length $|u_x|$, pointing with or against the x-arrowhead, depending on whether u_x is positive or negative. You next draw a vector parallel to the y-axis of length $|u_y|$, pointing with or against the y-arrowhead, depending on whether u_y is positive or negative. The vector **u** is the vector sum of the two vectors you just drew.

» WORKED EXAMPLE 6

A vector **v** has components $v_x = +6$ and $v_y = -3$ relative to the coordinate system shown in Figure 57. Sketch the vector **v** on the figure.

» SOLUTION 6

See Figure 58. ◇

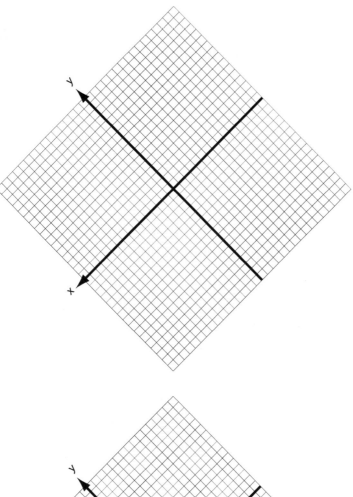

FIGURE 57.
The coordinate system to be used in Worked Example 6.

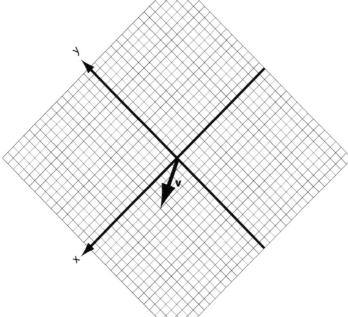

FIGURE 58.
Solution to Worked Example 6. The vector **v** with $v_x = +6$ and $v_y = -3$ relative to the coordinate system shown in Figure 57.

FIGURE 59.
Presenting vectors
in terms of their
x- and *y*-com-
ponents relative
to a coordinate
system.

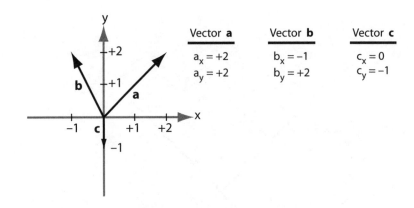

Vector **a**	Vector **b**	Vector **c**
$a_x = +2$	$b_x = -1$	$c_x = 0$
$a_y = +2$	$b_y = +2$	$c_y = -1$

FIGURE 59.
Presenting vectors
in terms of their
x- and *y*-com-
ponents relative
to a coordinate
system.

Presenting Vectors in Component Form

In Worked Example 6 we saw that a vector can be assembled from its *x*- and *y*-components. This means that if you had a specific vector in mind and you wanted to tell your friend what it was, one good way to do it would be to show your friend the vector's *x*- and *y*-components relative to a chosen coordinate system. See Figure 59.

In Figure 59, the *x*-component of vector **a** is +2. The + sign means that the *x*-shadow of **a** falls on the positive *x*-axis. The value 2 means that the *x*-shadow is 2 units long.

There is a handy notation for presenting vectors in component form. To present a vector in this way, all you do is write the vector in the form $\mathbf{v} = (v_x, v_y)$, where v_x and v_y are the *x*- and *y*-components of **v**. In this convention, the vectors in Figure 59 would be written as

Notice that vec-
tor **c** casts a
shadow of zero
length on the *x*-
axis. This means
that $c_x = 0$.

$$\mathbf{a} = (+2, +2), \tag{20}$$

$$\mathbf{b} = (-1, +2), \tag{21}$$

$$\mathbf{c} = (0, -1). \tag{22}$$

Using Components to Add and Subtract Vectors

Earlier we learned how to combine vectors visually using parallelogram addition, tip-to-tail addition, and tip-to-tip subtraction. Next we shall see that there is an easy way to add and subtract vectors quantitatively: by using their components. Given two vectors in component form, $\mathbf{u} = (u_x, u_y)$ and $\mathbf{v} = (v_x, v_y)$, their sum $\mathbf{u} + \mathbf{v}$ is the vector with components given by $(u_x + v_x, u_y + v_y)$. In other words,

$$(\mathbf{u} + \mathbf{v})_x = u_x + v_x \tag{23}$$

and

$$(\mathbf{u} + \mathbf{v})_y = u_y + v_y. \tag{24}$$

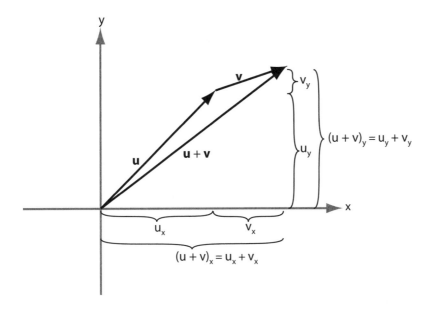

FIGURE 60.
Showing that tip-
to-tail addition and
component-wise
addition both give
the same result.

The x-component of the sum is the sum of the x-components. The y-component of the sum is the sum of the y-components.

Figure 60 shows that Equations 23 and 24 are equivalent to our visual method of tip-to-tail vector addition.

Likewise, when subtracting vectors, you can work quantitatively by subtracting components:

$$(\mathbf{u} - \mathbf{v})_x = u_x - v_x \tag{25}$$

and

$$(\mathbf{u} - \mathbf{v})_y = u_y - v_y. \tag{26}$$

» WORKED EXAMPLE 7

Find the components of the vector sum $\mathbf{b} + \mathbf{c}$, where $\mathbf{b}$ and $\mathbf{c}$ are given in Figure 59.

» SOLUTION 7

For convenience, we'll give the sum $\mathbf{b} + \mathbf{c}$ the name $\mathbf{s}$. To find the components of $\mathbf{s}$, we simply add the components of $\mathbf{b}$ and $\mathbf{c}$, keeping the x- and y-components separate:

$$s_x = b_x + c_x \tag{27}$$
$$= (-1) + (0) \tag{28}$$
$$= -1 \tag{29}$$

and

$$s_y = b_y + c_y \tag{30}$$

$$= (+2) + (-1) \tag{31}$$

$$= +1. \tag{32}$$

Thus, the components of **s** are $\mathbf{s} = (s_x, s_y) = (-1, +1)$. The vector **s** is shown in Figure 61. ◇

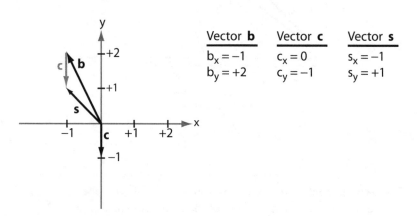

FIGURE 61.
Solution to Worked
Example 7. The sum
$\mathbf{s} = \mathbf{b} + \mathbf{c} = (-1, +1)$.

Vector **b**	Vector **c**	Vector **s**
$b_x = -1$	$c_x = 0$	$s_x = -1$
$b_y = +2$	$c_y = -1$	$s_y = +1$

How Scalar Multiplication Affects the Components of a Vector

When you multiply a vector **u** by a scalar such as 0.5, the new vector 0.5**u** will point in the same direction as **u**, but the new vector will be half as long as the original vector. To find the components of the new vector 0.5**u**, all you have to do is multiply each of the original components by 0.5. Figure 62 shows why this is so.

In general, given a scalar c and a vector **u** with components (u_x, u_y), the components of the vector $c\mathbf{u}$ are

$$c\mathbf{u} = (cu_x, cu_y). \tag{33}$$

If the scalar multiplication is expressed in the form of a division, the components are

$$\frac{\mathbf{u}}{a} = \left(\frac{u_x}{a}, \frac{u_y}{a} \right). \tag{34}$$

Vector Magnitude and Direction

Besides using the component form $\mathbf{v} = (v_x, v_y)$, another common way to present a vector is to state the magnitude of the vector and its direction. You might say, "The magnitude of the velocity

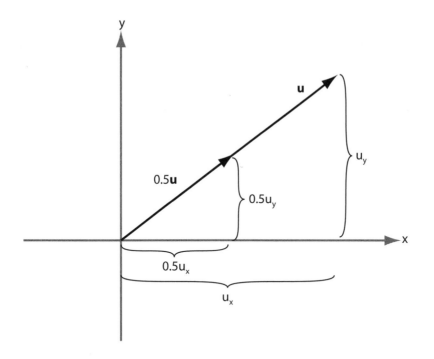

FIGURE 62.
Multiplying a vector by a scalar. This is the same thing as multiplying each component by the scalar.

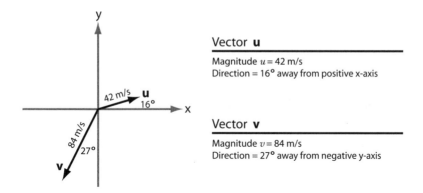

Vector u

Magnitude $u = 42$ m/s
Direction = 16° away from positive x-axis

Vector v

Magnitude $v = 84$ m/s
Direction = 27° away from negative y-axis

FIGURE 63.
Presenting vectors in magnitude-and-direction form.

is 42 m/s, and the direction is 16° above the positive x-axis." See Figure 63.

The text in Figure 63 uses a common notation in type for the magnitude of a vector **v**, namely v. In this book, when you see boldface, like **v**, you're looking at an arrow. When you see italics, like v, you're looking at an ordinary number.

The fact that there are two different ways to present vectors—either in terms of the x- and y-components or in terms of the magnitude and direction—means that next we have to talk about how to convert from one mode of presentation to the other.

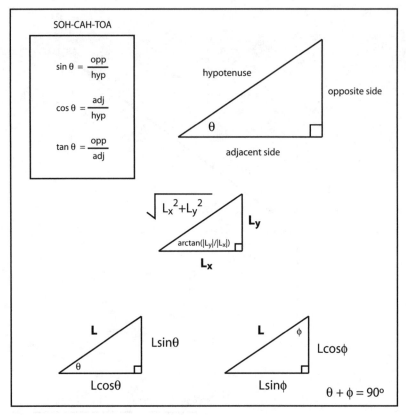

FIGURE 64. A summary of basic trigonometry facts.

Need a trig
refresher course?
For a start, make
a photocopy of
Figure 64 and
keep it handy.

Finding Magnitude and Direction
from the *x*- and *y*-Components

If someone gives you the *x*- and *y*-components of a vector, you can
use them to find the magnitude and direction of the vector. This
is easily done using simple trigonometry.

Given a vector **v**, with known components v_x and v_y, we can find
the magnitude and direction of **v**. The magnitude of **v** is given by
the Pythagorean Theorem:

(35)
$$v = \sqrt{v_x^2 + v_y^2}.$$

This is easy to see from Figure 65(a).

The direction angle θ of the vector **v** above the *x*-axis (see
Figure 65(b)) is given by

(36)
$$\theta = \arctan \frac{|v_y|}{|v_x|}.$$

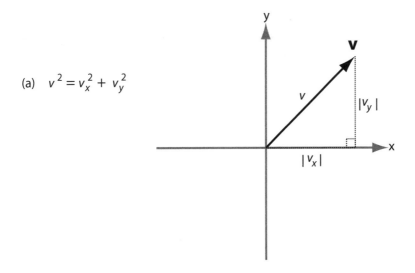

(a) $v^2 = v_x^2 + v_y^2$

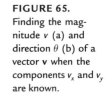

FIGURE 65.
Finding the magnitude v (a) and direction θ (b) of a vector **v** when the components v_x and v_y are known.

(b) $\theta = \arctan \dfrac{|\mathbf{v}_y|}{|\mathbf{v}_x|}$

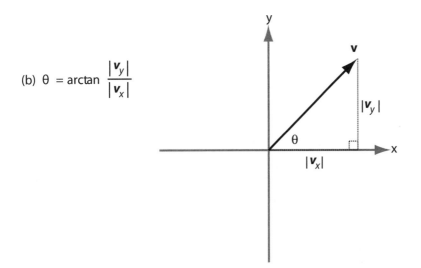

Finding x- and y-Components from the Magnitude and Direction

Conversely, if someone gives you the magnitude and direction of a vector, you can find the x- and y-components of the vector relative to your coordinate system. This is again easily done using simple trigonometry, as shown in the next Worked Example.

» WORKED EXAMPLE 8

Find the x- and y-components of a velocity vector $\mathbf{v}$ with magnitude 62 m/s and direction $\theta = 20°$ above the positive x-axis (Figure 66).

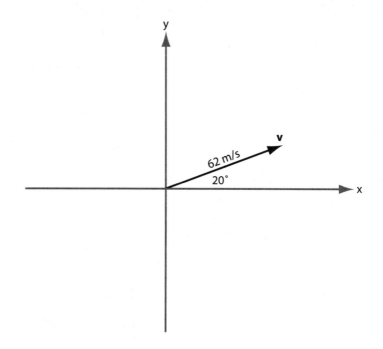

FIGURE 66.
Worked Example 8.
Finding the x- and y-components of a velocity vector $\mathbf{v}$.

» SOLUTION 8

First we visually resolve $\mathbf{v}$ into x- and y-components on the diagram, as shown in Figure 67.

Now we are looking at a right triangle, and the values of v_x and v_y can be found from trigonometry. But found *how*, exactly? Sometimes I find it helps students get started if we just pick a trig function and see what happens. Let's choose sine. Applying $\sin \theta = \frac{\text{opp}}{\text{hyp}}$ to the triangle in Figure 67, we have

$$\sin 20° = \frac{|v_y|}{62 \, \frac{\text{m}}{\text{s}}}. \tag{37}$$

Apparently this is going to give us the y-component, v_y. Multiplying both sides of the above equation by 62 m/s, we have

$$|v_y| = (62 \text{ m/s}) \sin 20° \tag{38}$$
$$= 21.2 \text{ m/s}.$$

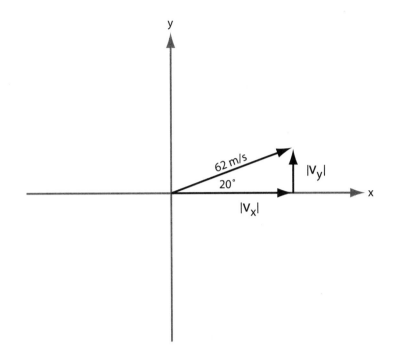

This is the absolute value of v_y. To find the sign of v_y, we have to look at the vector **v** in relation to the *y*-axis. The vector **v** points basically *with* the *y*-arrowhead, and therefore the *y*-component v_y should be *positive*. Hence we have

$$v_y = +21.2 \text{ m/s.} \tag{39}$$

With v_y in hand, we could proceed to find v_x in any number of ways. For one thing, we now know two out of the three legs of the triangle, so we could use the Pythagorean Theorem to find the third. But that approach is prone to round-off error, so instead let's just pick another trig function. I'll go with cosine. Applying $\cos \theta = \frac{\text{adj}}{\text{hyp}}$ to the triangle in Figure 67, we have

$$\cos 20° = \frac{|v_x|}{62 \frac{\text{m}}{\text{s}}}. \tag{40}$$

This is going to give us the *x*-component, as desired. Multiplying both sides by 62 m/s, we have

$$|v_x| = (62 \text{ m/s}) \cos 20°$$
$$= 58.3 \text{ m/s.} \tag{41}$$

To find the sign of v_x, we have to look at the vector **v** in relation to the *x*-axis. The vector **v** points basically *with* the *x*-arrowhead, so the x-component v_x should be *positive*. Hence we have

(42)
$$v_x = +58.3 \text{ m/s}.$$

We now have both components v_x and v_y, and we're done. ◇

Could you have selected the tangent function instead of the cosine function at the second step? Sure. Here's what would have happened. Applying $\tan \theta = \frac{\text{opp}}{\text{adj}}$ to the triangle in Figure 67, we have

(43)
$$\tan 20° = \frac{|v_y|}{|v_x|}$$

and

(44)
$$\tan 20° = \frac{21.2 \text{ m/s}}{|v_x|}.$$

Cross-multiplying and dividing both sides by $\tan 20°$, we have

(45)
$$|v_x| = \frac{21.2 \text{ m/s}}{\tan 20°} = 58.2 \text{ m/s}.$$

This approach suffers from round-off error, because it uses a rounded answer ($v_y = 21.2$ m/s) to find v_x, which we then round again. The cosine approach leading to Equation 41 was nicer because it used only "givens" to find v_x, minimizing the round-off error.

The velocity vector in this problem had a magnitude of 62 m/s. The physical meaning of this is easy to see: The object is moving with a speed of 62 meters per second. But what is the physical meaning of the two components we found, $v_x = 58.3$ m/s and $v_y = 21.2$ m/s? It is this: In one second, an object with the given velocity vector would move 58.3 meters horizontally and 21.2 meters vertically while covering a distance of 62 meters on its diagonal path. More on this in the next chapter.

Adding Vectors When They Are Given in the Form of Magnitude and Direction

You already know how to add vectors visually using the parallelogram law or the tip-to-tail method. You also know how to add vectors quantitatively by adding x- and y-components separately. But what if you are presented with two vectors in magnitude-and-direction form? Is there a way to add them in magnitude-

and-direction form? Unfortunately, no. The only way to add vectors quantitatively is to work with their components. Vectors can (and must) be added component-wise. So if someone gives you a pair of vectors in magnitude-and-direction form, in order to add them you will have to break them both into components, then add the components, and then assemble the result back into magnitude-and-direction form. Let's see how that looks.

» WORKED EXAMPLE 9

Vector **u** has magnitude 4 and points at an angle 45° above the positive x-axis. Vector **v** has magnitude 2 and points at an angle 30° above the negative x-axis. See Figure 68 for both. Find the magnitude and direction of **u** + **v**.

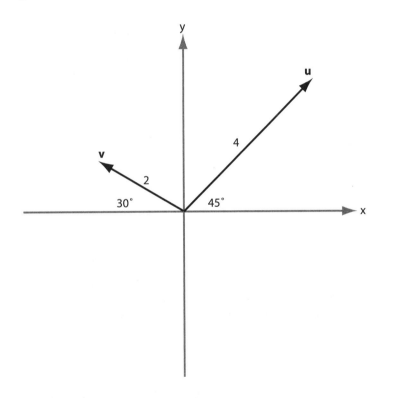

FIGURE 68.
Worked Example 9.
The vectors **u** and **v**
to be added.

» SOLUTION 9

To add the given vectors, we first have to find their x- and y-components. Then we'll add the x- and y-components separately to obtain the x- and y-components of the sum. And then we'll translate the components of the sum back into magnitude-and-direction terms.

No, there isn't a
faster way.

First we break **u** into components on the diagram, as shown in Figure 69.

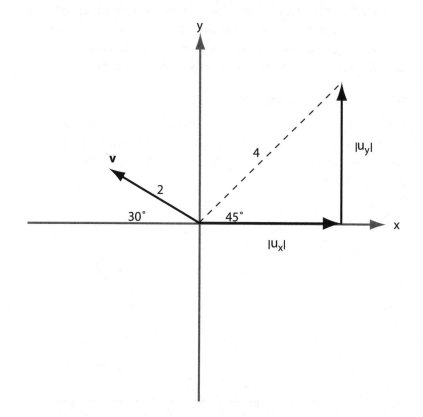

FIGURE 69.
Solving Worked Example 9. The vector u broken into components.

Once I break a vector into horizontal and vertical components, I like to erase the original "diagonal" vector to get it out of the way. After all, the two component vectors on the diagram are together equivalent to the original one.

The values of u_x and u_y are found from trigonometry as $u_x = +2.83$ and $u_y = +2.83$. (Try this yourself!)

Next we break **v** into components on the diagram, as shown in Figure 70. The values of v_x and v_y are found from trigonometry as $v_x = -1.73$ and $v_y = +1$. (Try this yourself!)

Did you notice the negative sign on v_x? Why is v_x negative?

Now we add the components. Let's give the sum $\mathbf{u} + \mathbf{v}$ the name **w**. Then the components of the sum vector **w** are

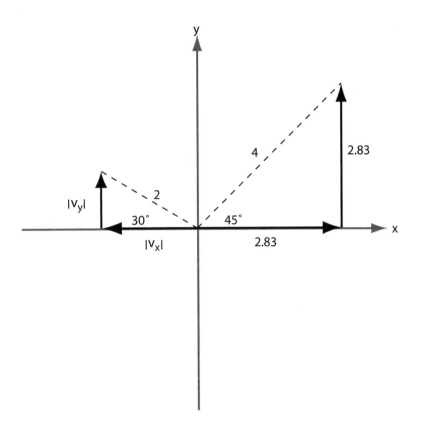

FIGURE 70.
Solving Worked
Example 9. The vec-
tor **v** broken into
components.

$$w_x = u_x + v_x \tag{46}$$

$$= (+2.83) + (-1.73) \tag{47}$$

$$= +1.1 \tag{48}$$

and

$$w_y = u_y + v_y \tag{49}$$

$$= (+2.83) + (+1) \tag{50}$$

$$= +3.83. \tag{51}$$

So the sum vector is given by

$$\mathbf{w} = (w_x, w_y) = (+1.1, +3.83). \tag{52}$$

The magnitude w is found from the Pythagorean Theorem as

$$w = \sqrt{w_x^2 + w_y^2} \tag{53}$$

$$= \sqrt{(1.1)^2 + (3.83)^2} \tag{54}$$

$$= 3.98 \tag{55}$$

and the angle of the vector sum above the x-axis is

(56)
$$\theta = \arctan \frac{|3.83|}{|1.1|} = 74°.$$

The final result is illustrated in Figure 71. ◇

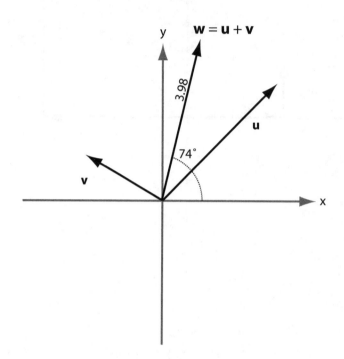

FIGURE 71.
Solution to Worked
Example 9. Adding
vectors **u** and **v** to
give **w**.

That's finally it! Our work with the mathematics of vectors is complete. Once you've done the Focused Problems, you'll be ready to apply what you know in the rest of the chapters to come.

Reader's Notes

Below is space for you to summarize key points from this chapter.

FOCUSED PROBLEMS :: CHAPTER 4

In the following problems, use a coordinate system in which x is positive to the right and y is positive upward.

1. For the vectors $\mathbf{F}_1$ and $\mathbf{F}_2$ shown in Figure 72, find the magnitude and direction of the vector sum $\mathbf{F}_{net} = \mathbf{F}_1 + \mathbf{F}_2$.

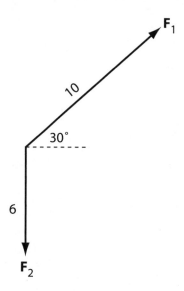

FIGURE 72.
Focused Problem 1.
Finding the magnitude and direction of the vector sum $\mathbf{F}_{net} = \mathbf{F}_1 + \mathbf{F}_2$.

2. For the vectors $\mathbf{F}_1$ and $\mathbf{F}_2$ shown in Figure 73, find the magnitude and direction of the vector sum $\mathbf{F}_{net} = \mathbf{F}_1 + \mathbf{F}_2$.

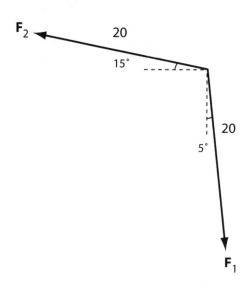

FIGURE 73.
Focused Problem 2.
Finding the magnitude and direction of the vector sum $\mathbf{F}_{net} = \mathbf{F}_1 + \mathbf{F}_2$.

3. For the vectors F_1 and F_2 shown in Figure 74, find the magnitude and direction of the vector sum $F_{net} = F_1 + F_2$.

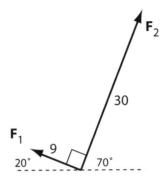

FIGURE 74.
Focused Problem 3.
Finding the magnitude and direction of the vector sum $F_{net} = F_1 + F_2$.

4. In Figure 75, find the magnitude F_1 of a vertically downward vector that will make the total vector sum $F_1 + F_2 + F_3$ equal to zero.

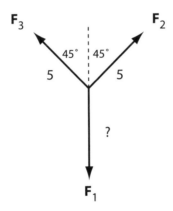

FIGURE 75.
Focused Problem 4.
Finding the magnitude F_1 of a vertically downward vector.

SOLUTIONS CAN BE FOUND ON PAGE 360

CHAPTER 5

Position and Velocity, Revisited

Chapter 3 introduced a few basic ideas about position and velocity:

- An object's position vector **r** tells how far from the origin the object is and in what direction.
- To draw the position vector, start with the tail at the origin and end with the tip at the location of the object.
- An object's velocity vector **v** tells how fast the object is moving and in what direction.
- To draw the velocity vector, start with the tail on the moving object.
- Draw the vector tangent to the trajectory.
- And make the vector long or short to reflect the speed of the object in qualitative terms.

In this chapter we'll apply our knowledge of vectors and components to work with position and velocity quantitatively. We'll also apply what we know about rates of change to gain a deeper understanding of how position and velocity relate to one another.

Interpreting the Components of Position Vectors

The x- and y-components of the position vector **r** have a simple meaning. We illustrate this with an example problem.

» WORKED EXAMPLE 1

At a particular instant of time, the position vector **r** of a taxicab has magnitude 68 m and direction $\theta = 30°$ away from the negative y-axis. See Figure 76. Find the components r_x and r_y of the taxicab's position vector at this time. What is the meaning of these components in simple terms?

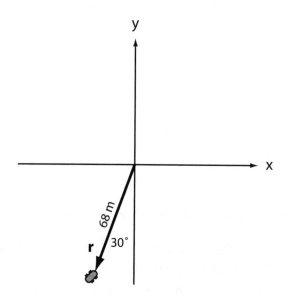

FIGURE 76.
Worked Example 1.
The position vector **r**
of a taxicab with
magnitude 68 m and
direction $\theta = 30°$
away from the
negative y-axis.

» SOLUTION 1

The x- and y-components of **r** are found using trigonometry as $r_x = -34$ m and $r_y = -58.9$ m. See Figure 77.

What is the physical meaning of these two components? In simple terms, r_x is the signed distance of the taxicab from the origin, measured in meters along the x-axis. And r_y is the signed distance of the taxicab from the origin, measured in meters along the y-axis.

Another way to say the same thing would be to say that r_x is the value of the cab's x-coordinate at any given time, and r_y is the value of the cab's y-coordinate at any given time. See Figure 78. ◇

Because r_x is the value of the x-coordinate at any given time, it is traditional to write r_x simply as x. This saves time, although it also obscures the fact that x refers to a component of the position vector **r**. Likewise, r_y is traditionally written as y. So when we're talking

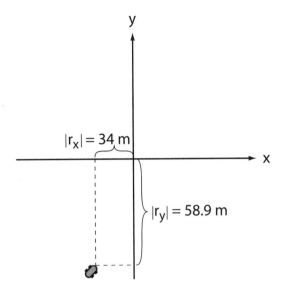

FIGURE 77.
Solution to Worked
Example 1. Interpret-
ing the components
r_x and r_y as signed
distances measured
along the principal
directions.

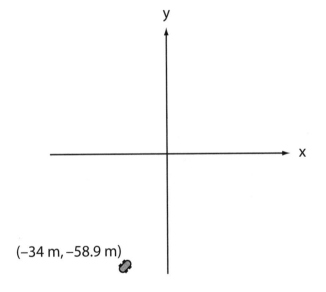

FIGURE 78.
Another solution
to Worked Exam-
ple 1. Another way to
interpret the compo-
nents r_x and r_y as the
(x, y) coordinates of
the taxicab.

about position vectors, the vector notation $\mathbf{r} = (r_x, r_y)$ could also
be written as $\mathbf{r} = (x, y)$.

The Relationship between Position and Velocity

It is important to understand position and velocity separately on
their own terms. That was the goal of Chapter 3. However, there is
also an important relationship between position and velocity. The

relationship was mentioned briefly at the end of Chapter 3, where we stated the formal definition of velocity:

(57)
$$\mathbf{v} = \frac{d}{dt}\mathbf{r}.$$

Read aloud, Equation 57 says that velocity is the rate of change of position.

Velocity is defined as the rate of change of position.

Our next goal is to understand Equation 57 by applying what we know about vectors and rates of change. In the process, we'll also learn how to solve quantitative problems involving position and velocity.

Using Two-Point Data to Estimate Velocity Vectors

Back in Chapter 2, we saw how to use two-point data to estimate rates of change for ordinary scalar quantities. (See Worked Example 4 and Focused Problem 6 from Chapter 2.) The basic rule was

(58)
$$\frac{d}{dt}Q \approx \frac{\Delta Q}{\Delta t} = \frac{Q_2 - Q_1}{t_2 - t_1}.$$

The quantities Q_1 and Q_2 in Equation 58 are ordinary real numbers. But in this chapter we are dealing with vectors that change over time. So what we would like to do is write a new rule, similar to Equation 58, that allows us to estimate the velocity vector $\mathbf{v} = \frac{d}{dt}\mathbf{r}$. In this case, the two-point data consist of two position vectors $\mathbf{r}_1$ and $\mathbf{r}_2$ at nearby times t_1 and t_2. See Figure 79.

By analogy with Equation 58, we can estimate the rate of change of $\mathbf{r}$ using two-point data as follows:

(59)
$$\frac{d}{dt}\mathbf{r} \approx \frac{\Delta \mathbf{r}}{\Delta t}$$
$$\mathbf{v} \approx \frac{\mathbf{r}_2 - \mathbf{r}_1}{t_2 - t_1}.$$

Note that Equation 59 implies that $\mathbf{v}$ is a scalar multiple of $\mathbf{r}_2 - \mathbf{r}_1$. (The scalar is $\frac{1}{t_2 - t_1}$.) That means that $\mathbf{v}$ points the same way as $\mathbf{r}_2 - \mathbf{r}_1$. Next, notice in Figure 79 the way tip-to-tip vector subtraction ensures that $\mathbf{r}_2 - \mathbf{r}_1$ points tangent to the trajectory. This guarantees that the velocity vector also points tangent to the trajectory, the way it should. Pretty cool!

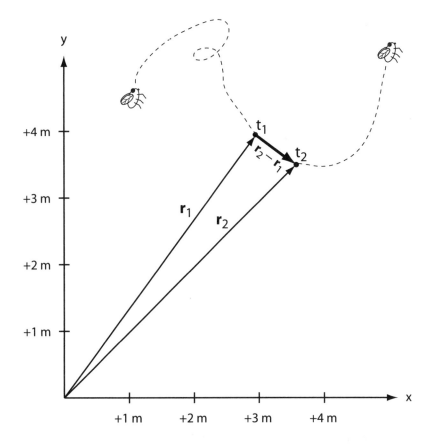

FIGURE 79.
Two-point data for estimating the velocity vector **v** for a fly.

Equation 59 gives us a great way to estimate the velocity vector based on the position vectors at two nearby instants of time. Let's see how this is done.

» WORKED EXAMPLE 2

Figure 80 shows two snapshots of a fly in motion. At the first instant of time shown, the fly's position vector is $\mathbf{r}_1 = (+3 \text{ m}, +4 \text{ m})$. One-tenth of a second later, the fly's new position vector is given by $\mathbf{r}_2 = (+3.2598 \text{ m}, +3.85 \text{ m})$. Estimate the fly's velocity vector during this interval of time. Give the answer in component form $\mathbf{v} = (v_x, v_y)$, and also give the speed and direction of the fly's motion.

» SOLUTION 2

We'll apply Equation 59 to the two-point data given in the problem. Remember that subtracting vectors quantitatively means subtracting their components (see Equations 25 and 26 in Chapter 4). So,

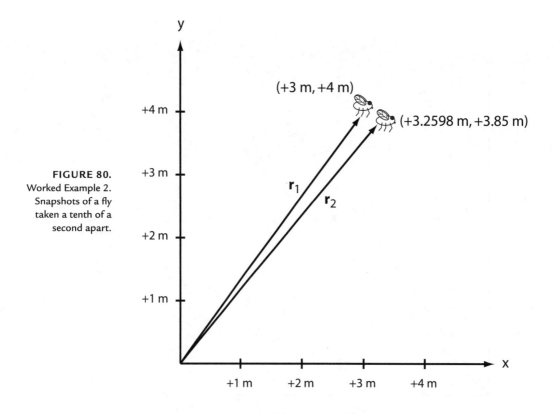

FIGURE 80.
Worked Example 2.
Snapshots of a fly
taken a tenth of a
second apart.

using the data given in the problem, Equation 59 becomes

$$\mathbf{v} \approx \frac{\mathbf{r}_2 - \mathbf{r}_1}{t_2 - t_1}$$

$$= \frac{(x_2, y_2) - (x_1, y_1)}{t_2 - t_1}$$

$$= \frac{(x_2 - x_1, \ y_2 - y_1)}{t_2 - t_1}$$

(60)

$$= \frac{(3.2598 \text{ m} - 3 \text{ m}, \ 3.85 \text{ m} - 4 \text{ m})}{0.1 \text{ s}}$$

$$= \frac{(0.2598 \text{ m}, -0.15 \text{ m})}{0.1 \text{ s}}$$

$$= \left(\frac{0.2598 \text{ m}}{0.1 \text{ s}}, \frac{-0.15 \text{ m}}{0.1 \text{ s}} \right)$$

$$(v_x, v_y) = \left(+2.598 \frac{\text{m}}{\text{s}}, -1.5 \frac{\text{m}}{\text{s}} \right).$$

We now have the x- and y-components of the fly's velocity vector. As for the speed and direction of motion, these are found from the components using trigonometry: $v = \sqrt{v_x^2 + v_y^2} = 3\,\text{m/s}$ at an angle $\arctan(1.5/2.598) = 30°$ below the positive x-direction. See Figure 81. $\diamond$

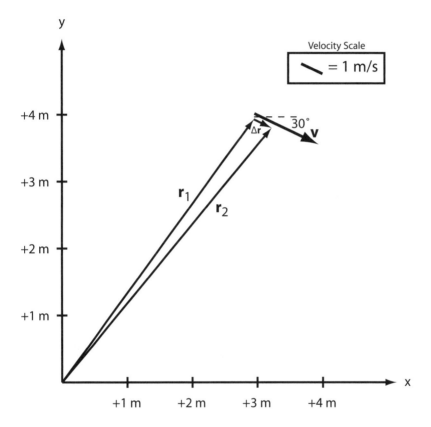

FIGURE 81.
Solution to Worked Example 2. The velocity vector of the fly during the tenth-of-a-second interval.

The length of a velocity vector when shown in a diagram does not relate in any visual way to the distance scale along the x- and y-axes. In Figure 81 I have shown a separate "velocity scale" so that the speed of the fly can be inferred visually from the length of its velocity vector. I like to think of velocity vectors as if they were drawn separately on a clear plastic transparency sheet and then overlaid on top of the diagram. The transparency sheet has its own separate magnitude scale, with units of meters per second.

Interpreting the Components of Velocity Vectors

Looking back at the third line from the top in Equation 60, we see that the x- and y-components of the velocity vector $\mathbf{v}$ are given by $v_x \approx \frac{x_2 - x_1}{t_2 - t_1}$ and $v_y \approx \frac{y_2 - y_1}{t_2 - t_1}$. In other words,

(61)
$$v_x \approx \frac{\Delta x}{\Delta t}$$
$$v_y \approx \frac{\Delta y}{\Delta t},$$

or, to be exact,

(62)
$$v_x = \frac{d}{dt}x$$
$$v_y = \frac{d}{dt}y.$$

These equations say that the x-velocity is the rate of change of the x-position, while the y-velocity is the rate of change of the y-position. This is the component-by-component version of our fundamental definition $\mathbf{v} = \frac{d}{dt}\mathbf{r}$.

To get a better sense of what the velocity components mean, let's return to the taxicab from Worked Example 1.

» WORKED EXAMPLE 3

The taxicab from Worked Example 1 has a velocity vector $\mathbf{v}$ with magnitude 10 m/s and direction $\theta = 60°$ away from the positive x-axis. See Figure 82. Find the components v_x and v_y of the taxicab's velocity vector at this time. What is the meaning of these components in simple terms?

» SOLUTION 3

The x- and y-components of $\mathbf{v}$ are found using trigonometry as $v_x = +5$ m/s and $v_y = +8.66$ m/s. What is the physical meaning of these numbers? It is this: If the taxicab moves with velocity vector $\mathbf{v}$ for one second, it will move 5 meters horizontally and 8.66 meters vertically as it covers a distance of 10 meters on a diagonal path. See Figure 83. ◇

---

v_x is the rate of change of x. And v_y is the rate of change of y.

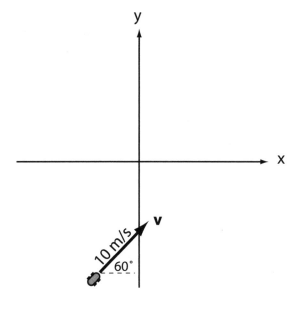

FIGURE 82.
Worked Example 3. The taxicab from Worked Example 1, with a velocity vector **v** with magnitude 10 m/s and direction $\theta = 60°$ away from the positive x-axis.

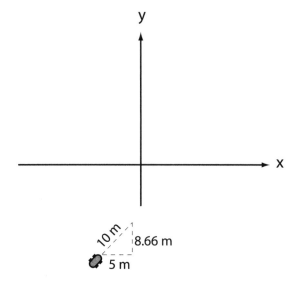

FIGURE 83.
Solution to Worked Example 3. Finding the components of the taxicab's velocity vector and deciding what they mean.

Reader's Notes

Below is space for you to summarize key points from this chapter.

FOCUSED PROBLEMS :: CHAPTER 5

1. A book lying on the floor has a position vector given by $\mathbf{r} = (-1\,\text{m}, +3\,\text{m})$. How far from the origin is this book?

2. At 3:00 p.m., an ocean liner was sighted 240 miles west of San Francisco and 120 miles north of San Francisco. At 3:15 p.m., the ocean liner was sighted 242 miles west of San Francisco and 119 miles north of San Francisco. Estimate the ocean liner's velocity vector during this time period, taking the x-direction to be east and the y-direction to be north. (Take the origin of the coordinates to be San Francisco.) Give the velocity vector in component form as $\mathbf{v} = (v_x, v_y)$, and also give the magnitude and direction of the velocity vector.

3. Now for a little warm-up on rates of change and proportional reasoning.
    a. There are 18 eggs in the refrigerator. You are using the eggs at a rate of 3 eggs per day. How many days will the eggs last?
    b. You are 18 meters from a brick wall. You are walking toward the wall at a rate of 3 meters per second. How long will it be before you crash into the wall?
    c. A ship is running at a speed of 18 knots. The ship is gaining speed at a rate of 3 knots per hour. How long will it take the ship to reach a speed of 30 knots?
    d. A car is driving at a speed of 36 m/s. The car is losing speed at a rate of 4 m/s per second. How long will it take the car to come to a stop?

4. While changing lanes on the highway, your velocity vector has magnitude 55 miles per hour and direction 3° to the right of the forward direction. (You are in the process of moving from the left-hand lane into the right-hand lane.)
    a. How long will it take you to complete the maneuver if the right-hand lane is 10 feet away?
    b. How far forward will you travel during this time?

5. Two boats are in a boat race. The finish line for the race is an east-west line stretched across the mouth of a channel.

   The lead boat is moving at a speed of 10 meters per second in a direction 20° north of east. The tip of the lead boat is located 100 meters south of the finish line.

   The tip of the trailing boat is located 110 meters south of the finish line. This boat is moving at a speed of 10 meters per second in a direction 22° north of west.

   Assuming that the two boats maintain constant velocities:

   a. Which boat will win the race? What will be the margin of victory (in seconds)?

   b. All other things being equal, what is the critical angle for the trailing boat that will divide winning from losing?

SOLUTIONS CAN BE FOUND ON PAGE 361

CHAPTER 6

Introducing Acceleration

You may think that you already know what *acceleration* means. Get ready for a new definition!

Here is what *acceleration* means:

$$
\text{Acceleration is} \begin{cases} \text{speeding up} \\ \text{slowing down} \\ \text{or turning.} \end{cases}
\tag{63}
$$

The word *acceleration* means something different in physics than it does in everyday speech. In everyday speech, *acceleration* means only "speeding up." But in physics, speeding up is just *one* of the meanings of the word *acceleration*.

Strangely enough, in physics *acceleration* can also mean slowing down! In common speech, slowing down is sometimes called "deceleration." But we shall never use this word. *Deceleration* is not a physics term. If we mean slowing down, we'll say "slowing down."

Finally, and even stranger still, in physics *acceleration* can also mean turning!

Acceleration is a big word that covers many situations that may seem unrelated at first. Let me emphasize five points:

1. If you are speeding up, you are accelerating.

2. If you are slowing down, you are accelerating.

3. If you are going at a constant speed but turning, you are accelerating.

4. If you are speeding up and turning at the same time, you are accelerating.

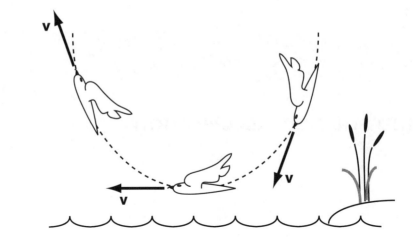

FIGURE 84.
The trajectory of a falcon that is turning at each instant of time shown.

5. If you are slowing down and turning at the same time, you are accelerating.

In fact, there are only two ways *not* to accelerate. One is to move in a perfectly straight line at a perfectly constant speed. The other is to stand still.

How to Tell When an Object Is Turning

Before we proceed with our discussion of acceleration, let me take a moment to clarify the word *turning*. By *turning* I mean moving with any degree of curvature in an object's trajectory. For example, if you are driving in a car and going around a bend, at that moment your trajectory is curved, and therefore you are turning.

Turning doesn't have to be a right-or-left thing. It can also be an up-or-down thing. Picture a falcon pulling out of a dive. The falcon in Figure 84 is turning at each moment shown because its trajectory is curved at each moment.

As another example, think about riding a bike. Is it possible for you to be "turning" even as you hold your handlebars perfectly straight? Sure—if you are riding on a hilly road! See Figure 85.

Students often have trouble identifying curvature in a trajectory. Identifying curvature is not as easy as it sounds. After all, if you "zoom in" on any curve far enough, it will look straight. So the trick is not to zoom in too much. When in doubt, draw a tangent line to the trajectory at the moment in question. If the trajectory falls away from the line to one side, the trajectory is curved at that point, and the object is turning at the moment in

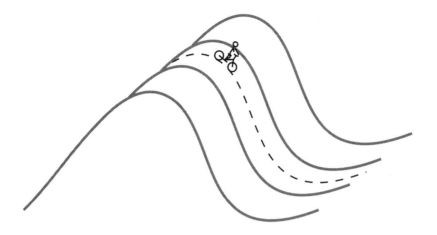

FIGURE 85.
The trajectory of a bicycle rider who is turning at the moment shown because her trajectory is curving in the vertical dimension. It doesn't matter that her trajectory would appear straight if viewed from directly overhead.

question. But if the trajectory "hugs" the tangent line and doesn't fall away—or if it falls away in different directions before and after the moment in question—the trajectory is straight, or at least momentarily straight, and the object is *not* turning at the moment in question.

An example is in order.

» WORKED EXAMPLE 1

A fly travels on the trajectory shown in Figure 86. For each moment of time indicated by a black dot, say whether the fly is turning at that moment or not.

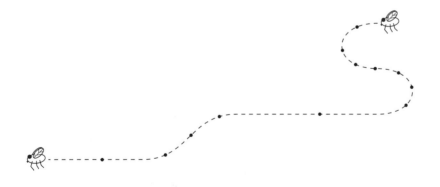

FIGURE 86.
Worked Example 1. The trajectory of a fly in flight, with moments in time represented by black dots.

» SOLUTION 1

In Figure 87, I have overlaid tangent lines on the trajectory from Figure 86. The fly is turning at moments 1, 2, 3, 5, 6, 7, 9, and 11. Observe how the trajectory falls away from the tangent line to one side at these moments.

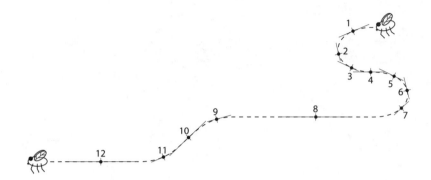

The fly is *not* turning at moments 4, 8, 10, and 12. At moments 8 and 12, the trajectory "hugs" the tangent line. At moments 4 and 10, the trajectory falls away from the tangent line to different sides just before and just after the chosen moment. ◇

Sketching the Acceleration Vector

As we'll see shortly, acceleration is one of those quantities for which an arrow representation makes a lot of sense. That is, acceleration is a vector quantity, which we denote in boldface by **a**.

We know how to sketch an object's position vector **r**: Just draw an arrow from the origin to the object. We also know how to sketch an object's velocity vector **v**: Just draw an arrow attached to the moving object, tangent to the object's path and roughly of a length that represents the object's speed. So how do we sketch an object's acceleration vector?

First of all, an object's acceleration vector is naturally drawn attached to the object—just like its velocity vector.

To draw a realistic acceleration vector at a given instant of time, you have to know how the object is moving at that instant: not only what its *velocity vector* is at that instant but also whether the object is speeding up, slowing down, or turning at that instant. If you know these things, drawing the object's acceleration vector is easy. Here are the possibilities; they are shown visually in Figure 88.

1. If the object is only speeding up (not turning at the same time), draw the acceleration vector *along* the velocity vector.

2. If the object is only slowing down (not turning at the same time), draw the acceleration vector *against* the velocity vector.

3. If the object is only turning (not speeding up or slowing down), draw the acceleration vector *perpendicular to* the

The turning of the falcon in Figure 84 will stand out more clearly to your eye if you sketch tangent lines to the falcon's trajectory at the three indicated moments of time.

Sketching acceleration vectors is one of the most important topics in this book. Fortunately, it's not hard to do. It just takes practice.

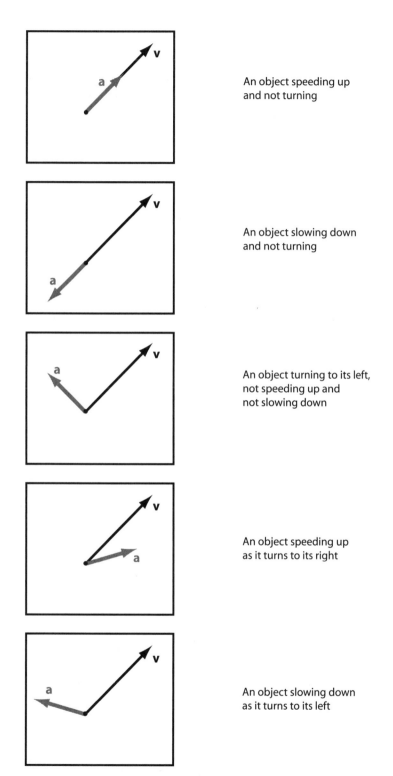

An object speeding up
and not turning

An object slowing down
and not turning

An object turning to its left,
not speeding up and
not slowing down

An object speeding up
as it turns to its right

An object slowing down
as it turns to its left

FIGURE 88. How to draw acceleration vectors.

velocity vector. The acceleration vector is like a "turn signal" in this case.

4. If the object is speeding up and turning at the same time, draw the acceleration vector *partly along and partly perpendicular to* the velocity vector.

5. If the object is slowing down and turning at the same time, draw the acceleration vector *partly against and partly perpendicular to* the velocity vector.

6. If the object is neither speeding up nor slowing down nor turning at the given moment, its acceleration vector is zero at that moment.

» WORKED EXAMPLE 2

A flying saucer orbits the earth in a circular path at a constant speed. Sketch the velocity and acceleration vectors of the flying saucer at four different times during its orbit.

» SOLUTION 2

Because the flying saucer is neither speeding up nor slowing down, we draw the acceleration vector perpendicular to the velocity vector. Make the acceleration vector indicate the direction of the turn. My sketches are shown in Figure 89. ◇

FIGURE 89.
Solution to
Worked
Example 2.
My sketches
of the velocity
and acceleration
vectors of the
flying saucer
described
in the text.

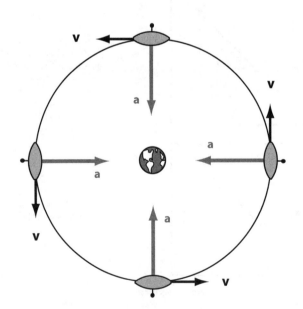

» WORKED EXAMPLE 3

Sketch velocity and acceleration vectors for the following three situations: (i) a car driving on a straight, flat road while braking; (ii) a rocket just after liftoff, when it is moving straight upward and speeding up; and (iii) a car speeding up as it pulls out of a driveway onto the main road.

» SOLUTION 3

My sketches are shown in Figure 90. (i) The car is not turning at all, so its acceleration vector must be purely along the velocity vector or purely against the velocity vector. The car is slowing down, so the acceleration vector is against the velocity vector. (ii) The rocket is not turning at all, so its acceleration vector must be purely along the velocity vector or purely against the velocity vector. The

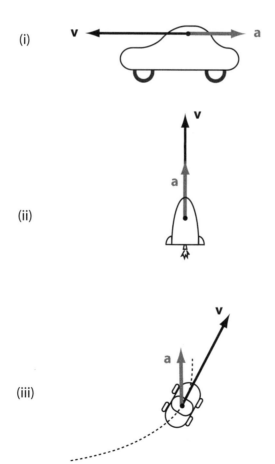

FIGURE 90.
Solution to Worked Example 3. My sketches of the velocity and acceleration vectors of the situations described in the text.

rocket is speeding up, so the acceleration vector is along the velocity vector. (iii) The car is speeding up and turning at the same time, so its acceleration vector points partly along the velocity vector and partly perpendicular to the velocity vector. ◇

»WORKED EXAMPLE 4

A clown is shot out of a cannon, with his trajectory shown in Figure 91.

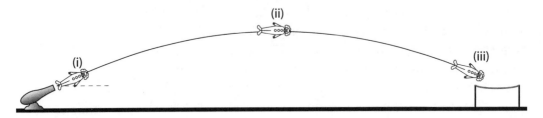

FIGURE 91.
Worked Example 4. The trajectory of a clown shot from a cannon and landing on a net.

Sketch the clown's velocity and acceleration vectors (i) just after he emerges from the cannon, (ii) at the peak of his trajectory, and (iii) just before he hits the net. Assume that the clown is slowing down on the way up and speeding up on the way down.

»SOLUTION 4

My sketches are shown in Figure 92.

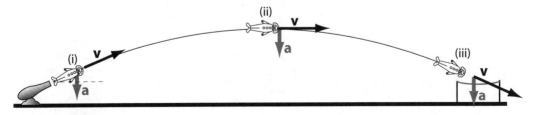

FIGURE 92.
Solution to Worked Example 4. My sketches of the velocity and acceleration vectors at three instants of time for the clown in Figure 91.

As the clown emerges from the cannon, we can see that he is turning because his trajectory is curved there. And we were told to assume that the clown is slowing down at this moment. Therefore the acceleration vector at this moment points partly perpendicular to the velocity vector and partly against it, as shown in Figure 92 (i). At the peak of his trajectory, the clown is turning; his trajectory is still curved. He has just finished slowing down, and he is about to start speeding up. Therefore, just at this moment, he is neither speeding up nor slowing down. So the acceleration vector is purely perpendicular to the velocity vector, as shown in Figure 92 (ii). At the end of the trajectory, he is turning and speeding up, so the acceleration vector points partly perpendicular to the velocity vector and partly along it, as shown in Figure 92 (iii).

We can therefore see, just by visualizing the motion, that the acceleration vectors must look *qualitatively* like those shown in Figure 92. Of course, just by thinking about it, you can't tell that all three acceleration vectors are actually the *same,* as is suggested by the figure. Nor can you tell that all three acceleration vectors are perfectly vertical, as is also suggested by the figure. But if we neglect air resistance, these things all turn out to be so, as we'll see in more detail later. ◇

The Vector Components a-Perp and a-Parallel

The acceleration vector **a** can be resolved into two perpendicular components: a component parallel to the velocity vector (either along it or against it) and a component perpendicular to the velocity vector. These components are shown in Figure 93. The component of the acceleration vector parallel to the velocity (either along it or against it) is denoted $a_{\parallel}$ and read aloud as "a-parallel." The

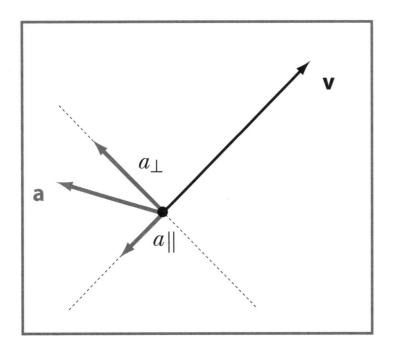

FIGURE 93.
Resolving **a** into components parallel and perpendicular to **v**. The object shown is in the process of slowing down ($a_{\parallel}$) and turning to its left ($a_{\perp}$).

component of the acceleration vector perpendicular to the velocity is denoted $a_\perp$ and read aloud as "a-perp."

I call $a_\parallel$ the "speeding-up part" of the acceleration vector **a** (or the "slowing-down part" if the object happens to be slowing down), and I call $a_\perp$ the "turning part" of the acceleration vector **a**.

» WORKED EXAMPLE 5

Stand up. (Really! Stand up and do this.) Now walk across the room at a constant speed, following an S-shaped path. Make your S as large as you can, and walk it pretty slowly, so that you can think about what you're doing. As you walk, use one arm to indicate the direction of your velocity vector at each instant of time, and use the other arm to indicate the direction of your acceleration vector at each instant of time.

» SOLUTION 5

Figure 94 shows a person using her arms to indicate the directions of her velocity vector **v** and her acceleration vector **a** as she follows an S-shaped path at a constant speed.

She keeps one arm pointed straight ahead of her; that arm represents her velocity vector. The velocity vector **v** always points tangent to her trajectory, straight out from her body as she walks.

Think of the acceleration vector in terms of its two basic components, the speeding-up or slowing-down part, $a_\parallel$, and the turning part, $a_\perp$. The person in the figure is walking at a constant speed, so there is no speeding-up or slowing-down part. That is, $a_\parallel = 0$. This means that the acceleration vector is always purely perpendicular to the velocity.

Think of $a_\perp$ as your "turn signal." It indicates which way your path is turning. The person in the figure uses her "acceleration arm" to point directly to her right or directly to her left, depending on which way she's turning at each instant.

Notice that in order to carry out this motion, you have to switch arms: during the first half of the S, your $a_\perp$ points to your left, but during the second half of the S, your $a_\perp$ points to your right.

From this it also follows that at the very center point of the S, your $a_\perp$ is zero. Moreover, because your $a_\parallel$ is *always* zero, this means

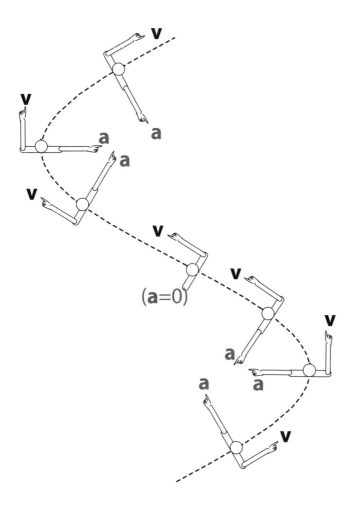

FIGURE 94.
Solution to Worked Example 5. Directions of the velocity and acceleration vectors for an S-shaped walk at constant speed.

that at the center of the S, your acceleration vector **a** vanishes altogether.

The center of the S is a special moment of time, and you can even feel this with your body as you carry out the motion. The center of the S feels to me like a "calm" moment when, for just a brief instant, I am coasting. ◇

A Tricky Case

I'll end this chapter with a tricky case that always comes up sooner or later.

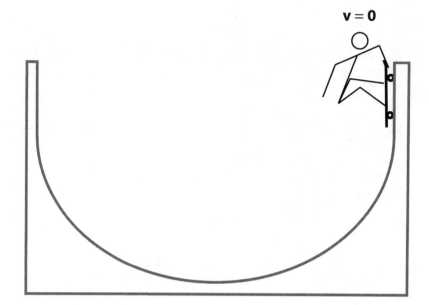

FIGURE 95.
A skateboarder just
as she comes to rest
at the top of a ramp.

You might want to skip to the end of the chapter and work on some of the Focused Problems, then come back to this topic.

The tricky case is that of an object *accelerating without moving*. Yes, this is possible!

Here's an example. Think of a skateboarder rolling back and forth on a U-shaped skateboard ramp (Figure 95).

Right at the moment when the skateboarder reaches her peak height, she is momentarily at rest. This presents us with a puzzle when it comes to sketching her acceleration vector, because our rules for sketching acceleration vectors tell us to draw **a** along **v**, **a** against **v**, **a** perpendicular to **v**, or some combination of these. If there's no **v** at all, how can you do this? Clearly, we need another rule to cover the case when the object is momentarily at rest.

For the case of the skateboarder, it turns out—and we'll discuss this in a moment—that the acceleration vector at the peak of her trajectory is as shown in Figure 96. The acceleration vector at this moment points straight down.

There is a good way to understand the acceleration vector in this case, and in fact it's a good way to view acceleration in general.

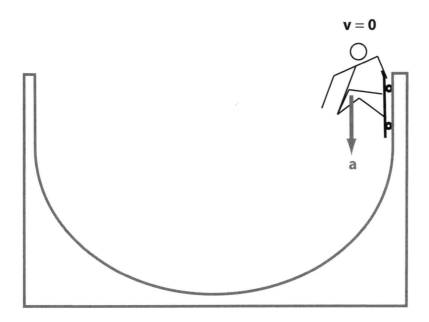

FIGURE 96.
The acceleration vector for the skate-boarder at the peak of her trajectory.

Think of the acceleration vector as an indicator of what is *about to happen* to the velocity vector. When the skateboarder is at the peak of her trajectory, she is not moving—but she's *about to be moving*. There may be no velocity vector at the moment in question, but there is *about to be one*—and it will be pointing downward once it appears. Because **a** tells what's just about to happen to **v**, it follows that **a** points downward right then.

See Focused Problem 5 of Chapter 7 for more on the acceleration vector as an indicator of what is "about to happen" to the velocity vector.

Figure 96 shows that it is possible to be accelerating without moving. This may seem crazy at first, but bear in mind that it can happen only for isolated single moments of time. At these isolated moments, the acceleration vector points in the direction in which the velocity vector is *about to point*.

Reader's Notes

Below is space for you to summarize key points from this chapter.

FOCUSED PROBLEMS :: CHAPTER 6

1. Driver to mechanic: "Can you take a look at my car? I think there's something wrong with my accelerator."

    Mechanic: "Which one?"

    What's the joke here? How many "accelerators" does a car have?

2. Give examples of motions in which the velocity vector and the acceleration vector of a moving object point (a) in the same direction, (b) in opposite directions, (c) in perpendicular directions, and (d) at an acute angle (an angle less than 90°).

3. Draw cartoons of the following, and sketch the velocity and acceleration vectors in each case:

    a. A tiny tot swinging in a swing at the moment when he is at the bottom of his trajectory.

    b. Same tot a few moments after reaching the bottom of his trajectory.

    c. Same tot a few moments before reaching the bottom of his trajectory.

4. Have you ever thought about what would happen if you dropped a penny from the Empire State Building? Draw a cartoon of this situation. Show the penny at three instants of time during its fall. For each instant, sketch the velocity vector and the acceleration vector of the penny.

5. A day in the life of an elevator. Draw pictures of the following, showing velocity and acceleration vectors in each case.

    a. The elevator at a moment when it is moving upward and speeding up.

    b. Same elevator at a moment when it is moving upward at a constant speed.

    c. Same elevator, at a moment when it is slowing to a stop (still moving upward).

    d. Same elevator at a moment when it is moving downward and speeding up.

e. Same elevator at a moment when it is moving downward at a constant speed.

f. Same elevator at a moment when it is slowing to a stop (still moving downward).

6. Sketch the acceleration vector of the fly in Figure 86 at each of the indicated moments of time. Don't worry about the magnitude of the acceleration vector, just pay attention to the direction. First assume that the fly is always moving at a constant speed. Then redraw your sketches assuming that the fly is always speeding up.

SOLUTIONS CAN BE FOUND ON PAGE 362

CHAPTER 7

Acceleration as a Rate of Change

The Relationship between Velocity and Acceleration

Imagine cruising along in your car with a constant, unchanging velocity vector **v**. Your car is just humming along, following a straight, flat road, maintaining a constant speed. Bored by the monotony, at risk of falling asleep at the wheel, you decide that you want to *change* your velocity vector. What are some ways you could do it?

1. Well, you could speed up. That would lengthen your velocity vector.

2. You could also slow down. That would shorten your velocity vector.

3. Finally, you could turn. That would rotate your velocity vector.

Well now, that's interesting: The three ways of changing your velocity vector are none other than the three ways to accelerate: speeding up, slowing down, and turning! Accelerating *means* changing your velocity vector.

Given this observation, it may come as no surprise to learn that acceleration is defined mathematically as the rate of change of the velocity vector:

$$\mathbf{a} = \frac{d}{dt}\mathbf{v}. \tag{64}$$

Acceleration is the rate of change of velocity. Acceleration is *not* the rate of change of speed. The rate of change of speed is actually $a_\parallel$, which is only one component of the acceleration vector—the speeding-up or slowing-down part.

The goal of this chapter is to gain a solid conceptual understanding of Equation 64 and, in the process, to learn how to solve quantitative problems involving acceleration.

Using Two-Point Data to Estimate Acceleration Vectors

Back in Chapter 5, we saw how to use two nearby position vectors to estimate the velocity vector. (See Worked Example 2 and Focused Problem 2 from Chapter 5.) The basic rule was

$$(65) \qquad \mathbf{v} = \frac{d}{dt}\mathbf{r} \approx \frac{\Delta \mathbf{r}}{\Delta t} = \frac{\mathbf{r}_2 - \mathbf{r}_1}{t_2 - t_1}.$$

Equations are complete sentences that are meant to be read aloud. Equation 65 is to be read something like this: "Velocity is the rate of change of position, which is approximately the change in the position vector over time, which is the difference in nearby position vectors over the difference in times." Notice that my sentence used only English words, no symbols like "r-two" or "vee." If you can't recite an equation as a complete sentence, using words instead of symbols, you haven't yet understood it.

A very similar procedure can be used again to estimate acceleration based on nearby velocities. See Figure 97(a).

Using the two-point velocity data shown in Figure 97(a), we can estimate the acceleration vector (Figure 97(b)) as

$$(66) \qquad \mathbf{a} = \frac{d}{dt}\mathbf{v} \approx \frac{\Delta \mathbf{v}}{\Delta t} = \frac{\mathbf{v}_2 - \mathbf{v}_1}{t_2 - t_1}.$$

> Read the equation aloud ... in a complete sentence ... using words, not symbols.

Equation 66 implies that $\mathbf{a}$ is a scalar multiple of $\mathbf{v}_2 - \mathbf{v}_1$. (The scalar is $\frac{1}{t_2-t_1}$.) That means that $\mathbf{a}$ points the same way as $\mathbf{v}_2 - \mathbf{v}_1$, as shown in Figure 97(b). Note the way that in Figure 97(c) tip-to-tip vector subtraction causes $\mathbf{a}$ to point in the correct direction to reflect the slowing down and turning of the fly. Pretty cool!

Equation 66 gives us a great way to estimate the acceleration vector from two nearby velocity vectors. Let's see how this is done.

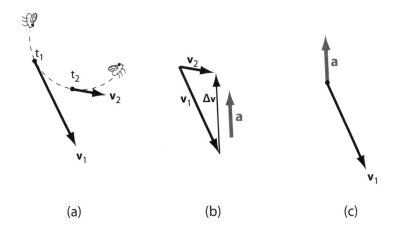

FIGURE 97.
(a) Velocity vectors for a fly at times t_1 and t_2. Observe, by looking at v_1 and v_2, that the fly is slowing down and turning.
(b) The acceleration vector **a**, given approximately by $\frac{v_2 - v_1}{t_2 - t_1}$.
(c) The acceleration vector pointing in the proper direction to represent slowing down and turning.

» WORKED EXAMPLE 1

Figure 98 shows a fly in motion. At time $t_1 = 0$, the fly's position vector is $\mathbf{r}_1 = (+3 \text{ m}, +4 \text{ m})$. At time $t_2 = 0.1$ s, the fly's new position vector is given by $\mathbf{r}_2 = (+3.2598 \text{ m}, +3.85 \text{ m})$. At time $t_3 = 0.2$ s, the fly's new position vector is given by $\mathbf{r}_3 = (+3.5196 \text{ m}, +3.85 \text{ m})$.

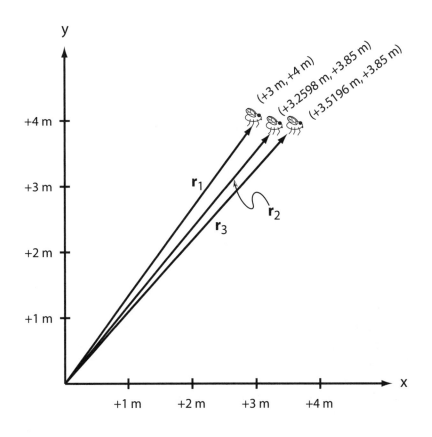

FIGURE 98.
Worked Example 1. Three snapshots of a fly taken at tenth-of-a-second intervals.

Estimate the fly's two (different) velocity vectors during the two given intervals of time. Use the two velocity vectors to estimate the fly's acceleration vector during the time span of the problem. Give the answer in component form, $\mathbf{a} = (a_x, a_y)$, and also give the magnitude and direction of the fly's acceleration vector.

» SOLUTION 1

First we'll use Equation 65, with $\mathbf{r}_1$ and $\mathbf{r}_2$, to find the velocity vector $\mathbf{v}_{1\to2}$ during the first time interval. Actually, we did this very same computation in Worked Example 2 of Chapter 5; the result was

$$(67) \qquad \mathbf{v}_{1\to2} = \left(+2.598\,\frac{m}{s}, -1.5\,\frac{m}{s}\right).$$

Next we'll use Equation 65 again, this time with $\mathbf{r}_2$ and $\mathbf{r}_3$, to find the velocity vector $\mathbf{v}_{2\to3}$ during the second time interval:

$$(68) \qquad \begin{aligned} \mathbf{v}_{2\to3} &\approx \frac{\mathbf{r}_3 - \mathbf{r}_2}{t_3 - t_2} \\[2mm] &= \frac{(x_3, y_3) - (x_2, y_2)}{t_3 - t_2} \\[2mm] &= \frac{(x_3 - x_2,\ y_3 - y_2)}{t_3 - t_2} \\[2mm] &= \frac{(3.5196\text{ m} - 3.2598\text{ m},\ 3.85\text{ m} - 3.85\text{ m})}{0.2\text{ s} - 0.1\text{ s}} \\[2mm] &= \frac{(+0.2598\text{ m}, 0\text{ m})}{0.1\text{ s}} \\[2mm] &= \left(\frac{+0.2598\text{ m}}{0.1\text{ s}}, \frac{0\text{ m}}{0.1\text{ s}}\right) \\[2mm] \mathbf{v}_{2\to3} &= \left(+2.598\frac{m}{s}, 0\frac{m}{s}\right). \end{aligned}$$

Two observations on the results so far: (1) During the time interval $t_2 \to t_3$, the fly moved purely horizontally (the y-position remained constant at $+3.85$ m). As a result, the y-component of $\mathbf{v}_{2\to3}$ came out to be zero. (2) The fly maintained a constant horizontal velocity throughout the problem, moving 0.2598 m along the x-axis during each of the 0.1-second intervals. This is why both $\mathbf{v}_{1\to2}$ and $\mathbf{v}_{2\to3}$ have x-components equal to $2.598\frac{m}{s}$.

Finally, we can use Equation 66, with our two velocity vectors $\mathbf{v}_{1\to2}$ and $\mathbf{v}_{2\to3}$, to find the acceleration vector $\mathbf{a}$ during the time span of the problem. For the purposes of figuring out what Δt is, we can say that the velocity $\mathbf{v}_{1\to2}$ occurred halfway between times t_1 and t_2, that is, at time $t_{12} = 0.05$ s. Likewise, we can say that the velocity $\mathbf{v}_{2\to3}$ occurred halfway between times t_2 and t_3, that is, at time $t_{23} = 0.15$ s. Now Equation 66 gives

$$\mathbf{a} \approx \frac{\mathbf{v}_{2\to3} - \mathbf{v}_{1\to2}}{t_{23} - t_{12}}$$

$$= \frac{(v_{2\to3,x}, v_{2\to3,y}) - (v_{1\to2,x}, v_{1\to2,y})}{t_{23} - t_{12}}$$

$$= \frac{(v_{2\to3,x} - v_{1\to2,x}, \quad v_{2\to3,y} - v_{1\to2,y})}{t_{23} - t_{12}}$$

$$= \frac{\left(2.598\frac{\text{m}}{\text{s}} - 2.598\frac{\text{m}}{\text{s}}, \quad 0\frac{\text{m}}{\text{s}} - -1.5\frac{\text{m}}{\text{s}}\right)}{0.15 \text{ s} - 0.05 \text{ s}} \tag{69}$$

$$= \frac{(0\frac{\text{m}}{\text{s}}, +1.5\frac{\text{m}}{\text{s}})}{0.1 \text{ s}}$$

$$= \left(\frac{0\frac{\text{m}}{\text{s}}}{0.1 \text{ s}}, \frac{+1.5\frac{\text{m}}{\text{s}}}{0.1 \text{ s}}\right)$$

$$(a_x, a_y) = \left(0\frac{\frac{\text{m}}{\text{s}}}{\text{s}}, +15\frac{\frac{\text{m}}{\text{s}}}{\text{s}}\right).$$

We now have the x- and y-components of the fly's approximate acceleration vector during the time span shown in Figure 98. As for the magnitude and direction of the acceleration vector, these are found from the components using trigonometry. The magnitude is $a = \sqrt{a_x^2 + a_y^2} = 15\frac{\text{m}}{\text{s}}$, and $\mathbf{a}$ points in the positive y-direction. See Figure 99. ◇

The magnitude of the acceleration vector in Figure 99 is equal to 15 meters per second per second. The length of an acceleration vector when shown in a diagram does not relate in any visual way to the distance scale along the x- and y-axes. I have included separate velocity and acceleration scales in Figure 99 so that the speed and the magnitude of the acceleration can be inferred visually from the length of the $\mathbf{v}$ and $\mathbf{a}$ vectors. I like to think of position, velocity, and acceleration vectors as if they were drawn

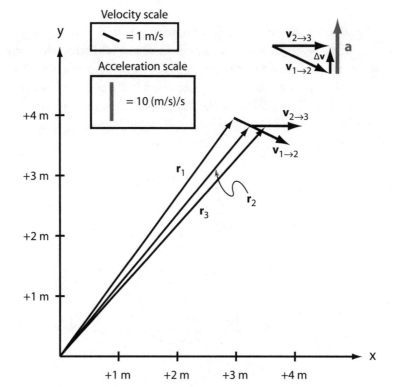

FIGURE 99.
Solution to Worked
Example 1. The
acceleration vector of
the fly during the
time frame of the
problem.

separately on clear plastic transparency sheets, then overlaid on the
diagram one after another. Each transparency sheet has its own separate
magnitude scale, measured in the appropriate units (meters, meters per
second, or meters per second per second).

Interpreting the Components of Acceleration Vectors

From Equation 66, it follows that the x- and y-components of the
acceleration vector **a** are given by

$$a_x \approx \frac{\Delta v_x}{\Delta t}$$

(70)

$$a_y \approx \frac{\Delta v_y}{\Delta t}$$

or, to be exact,

$$a_x = \frac{d}{dt} v_x$$

(71)

$$a_y = \frac{d}{dt} v_y.$$

The x-acceleration is the rate of change of the x-velocity, while the y-acceleration is the rate of change of the y-velocity. This is the component-by-component version of our fundamental definition $\mathbf{a} = \frac{d}{dt}\mathbf{v}$.

As we saw in Worked Example 1, the units of acceleration are $\frac{\frac{m}{s}}{s}$. This is usually written as m/s^2. We'll follow this convention, although $\frac{\frac{m}{s}}{s}$ makes it clearer that acceleration measures the rate of change of a quantity measured in $\frac{m}{s}$.

In brief: a_x is the rate of change of v_x, and a_y is the rate of change of v_y.

More about the x- and y-Components of the Acceleration Vector

As we saw in Chapter 6, one way to decompose the acceleration vector is by drawing the speeding-up or slowing-down component $a_\parallel$ along the velocity vector and the turning component $a_\perp$ perpendicular to the velocity vector. The previous examples show that you can also break the acceleration vector into x- and y-components a_x and a_y relative to a coordinate system. The components a_x and a_y will generally not tell you directly whether an object is speeding up, slowing down, or turning. (They will tell you this only if one of your coordinate axes happens to lie along the velocity vector. Then a_x and a_y will *be* $a_\parallel$ and $a_\perp$, or vice versa.) However, the coordinate breakdown of the acceleration vector is often useful for other reasons, as we'll see when we begin to analyze forces. So let's get some more practice with a_x and a_y.

» WORKED EXAMPLE 2

A fly has an acceleration vector $\mathbf{a}$ of magnitude $12\ m/s^2$, pointing at an angle $60°$ away from the negative y-axis, as shown in Figure 100. Find a_x and a_y, the x- and y-components of the fly's acceleration vector.

» SOLUTION 2

First we break $\mathbf{a}$ into x- and y-components on the diagram, as shown in Figure 101.

Now the values of a_x and a_y are found from trigonometry. To do this, let's just pick a trig function and see what happens. Let's choose sine. Applying $\sin\theta = \frac{\text{opp}}{\text{hyp}}$ to the triangle in Figure 101, we have

$$\sin 60° = \frac{|a_x|}{12\ m/s^2}. \tag{72}$$

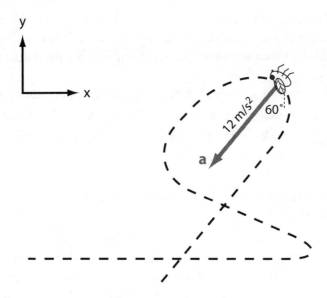

FIGURE 100.
Worked Example 2.
A fly with its accel-
eration vector **a**.

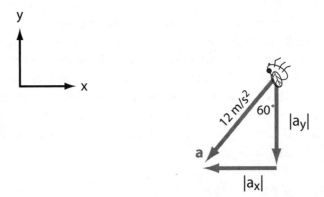

FIGURE 101.
Solving Worked
Example 2 by
breaking **a** into *x*-
and *y*-components.

Apparently this is going to give us the *x*-component. Multiplying both sides of Equation 72 by 12 m/s², we have

$$\left|a_x\right| = (12 \text{ m/s}^2)\sin 60°$$

$$= 10.4 \text{ m/s}^2.$$

(73)

To get the sign of a_x, we have to look at the vector **a** in relation to the *x*-axis. The vector **a** points basically *against* the *x*-arrowhead, so the *x*-component a_x should be *negative*. Hence we have

$$a_x = -10.4 \text{ m/s}^2.$$

(74)

With a_x in hand, we could proceed to find a_y in any of a number of ways. For one thing, we now know two out of the three legs of the triangle, so we could use the Pythagorean Theorem to find the third. But that approach is prone to round-off error, so instead let's just pick another trig function. I'll go with cosine. Applying $\cos\theta = \frac{\text{adj}}{\text{hyp}}$ to the triangle in Figure 101, we have

$$\cos 60° = \frac{|a_y|}{12 \text{ m/s}^2}. \tag{75}$$

(This is going to give us the y-component, as desired.) Multiplying both sides by 12 m/s^2, we have

$$|a_y| = (12 \text{ m/s}^2) \cos 60°$$
$$= 6.0 \text{ m/s}^2. \tag{76}$$

To get the sign of a_y, we have to look at the vector **a** in relation to the y-axis. The vector **a** points basically *against* the y-arrowhead, so the y-component a_y should be *negative*. Hence we have

$$a_y = -6.0 \text{ m/s}^2. \tag{77}$$

Could you have selected the tangent function instead of the cosine function? Sure. Here's what would have happened. Applying $\tan\theta = \frac{\text{opp}}{\text{adj}}$ to the triangle in Figure 101, we have

$$\tan 60° = \frac{|a_x|}{|a_y|} \tag{78}$$

$$\tan 60° = \frac{10.4 \text{ m/s}^2}{|a_y|}. \tag{79}$$

Cross-multiplying and dividing both sides by $\tan 60°$, we have

$$|a_y| = \frac{12 \text{ m/s}^2}{\tan 60°}$$
$$= 6.0 \text{ m/s}^2. \tag{80}$$

This time we got lucky and there was no round-off error, even though we used a rounded answer (a_x) to get a_y, which we then rounded again. ◇

What is the physical meaning of our answers $a_x = -10.4 \text{ m/s}^2$ and $a_y = -6.0 \text{ m/s}^2$?

- The physical meaning of $a_x = -10.4 \text{ m/s}^2$ is that, at the moment shown in the figure, the fly's horizontal velocity component, v_x, is decreasing at a rate of 10.4 m/s every second.

- The physical meaning of $a_y = -6.0 \, \text{m/s}^2$ is that, at the moment shown in the figure, the fly's vertical velocity component, v_y, is decreasing at a rate of 6 m/s every second.

These facts are perhaps not so illuminating in themselves. For example, they don't tell us whether the fly is speeding up, slowing down, or turning at the moment of time in question. To know that, we'd need to resolve the acceleration vector not into x-and y-components a_x and a_y but into parallel and perpendicular components $a_\parallel$ and $a_\perp$ relative to the velocity vector itself. Indeed, a simple glance at Figure 100 shows that at this moment of time, the fly is turning but neither speeding up nor slowing down. (How could I tell?)

Although the coordinate components a_x and a_y may not tell us directly about the character of the motion, the way $a_\parallel$ and $a_\perp$ do, they at least have the virtue of allowing us to deal with motion "one component at a time." This will turn out to be extremely useful later on. Worked Example 3 gives us a first look at this kind of analysis.

»WORKED EXAMPLE 3

Figure 102 shows the trajectory of the clown from Worked Example 4 in Chapter 6. The clown's velocity and acceleration vectors are shown at three specific instants of time. On the given axes, sketch v_x, v_y, a_x, a_y, and v as functions of time during the whole of the clown's flight.

»SOLUTION 3

There are many ways to begin this problem. I'll start by noticing that in our given scenario, the acceleration vector **a** is *always vertical*. That means that the x-component is always zero: $a_x = 0$. So the graph of a_x versus time is easy to draw (Figure 103).

Next I want to use what we know about rates of change. Because a_x is always zero and because a_x is the rate of change of v_x, the rate of change of v_x is zero. But that just means that v_x is constant over time. I've shown this in Figure 104.

This is kind of interesting: To the extent that the acceleration vector does indeed point purely vertically, it means that the horizontal velocity component remains constant. The clown tracks across the scene horizontally at a constant rate. A dog trotting

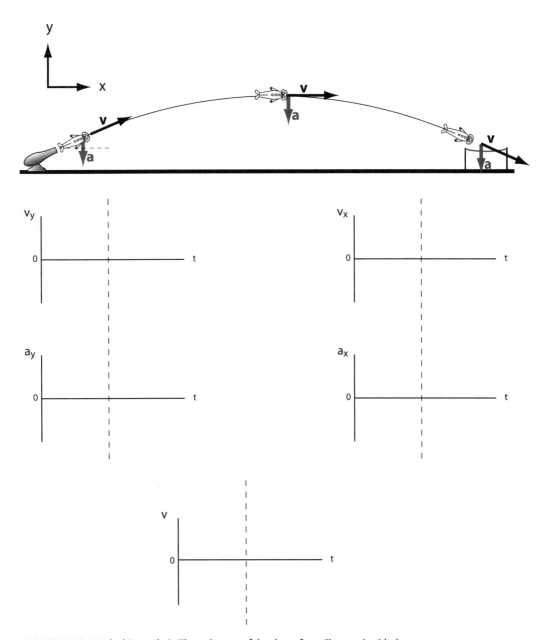

FIGURE 102. Worked Example 3. The trajectory of the clown from Chapter 6, with the clown's velocity and acceleration vectors shown at three specific instants of time.

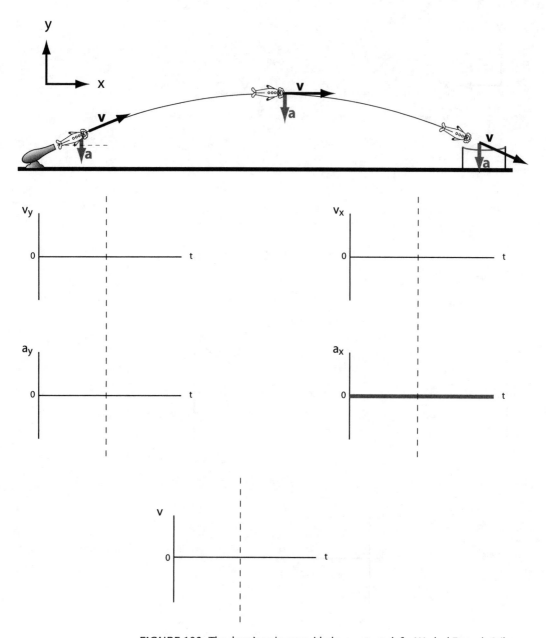

FIGURE 103. The clown's trajectory with the $a_x - t$ graph for Worked Example 3 (lower right-hand corner). Look back and forth between the $a_x - t$ graph and the acceleration vectors in the diagram until you can see how the graph reflects the behavior of the acceleration vector over time.

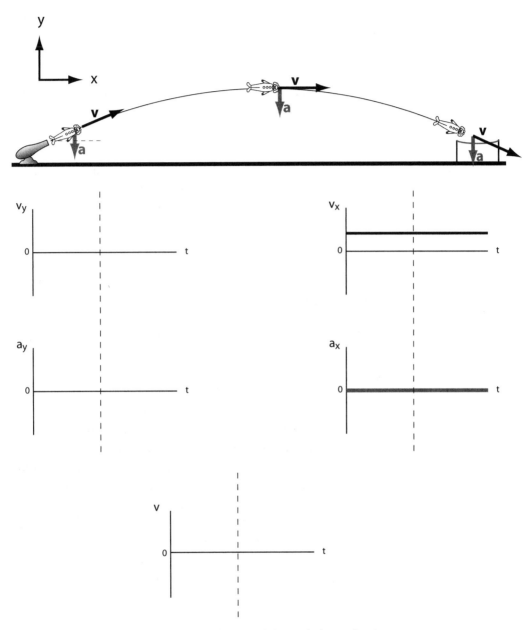

FIGURE 104. The clown's trajectory with the $v_x - t$ graph for Worked Example 3 (upper right-hand corner). Look back and forth between the $a_x - t$ graph and the $v_x - t$ graph until you can see that the $a_x - t$ graph shows the rate of change of the $v_x - t$ graph.

along the ground at a constant speed underneath the clown would always remain within the clown's shadow.

I knew to make the constant value of v_x positive because the velocity vector always points *with* the arrowhead on the x-axis.

Now let's think about y-components. Looking at the acceleration vectors shown in the original diagram, we see that the vertical component a_y appears to maintain a *constant negative value over time*. (Note that the **a** vectors point *against* our y-axis.) The $a_y - t$ graph in Figure 105 records this observation.

We can use rate-of-change reasoning once again to deduce the $v_y - t$ graph from the $a_y - t$ graph. The $a_y - t$ graph has a *constant negative value*. Because a_y is the rate of change of v_y, this means that the $v_y - t$ graph must have a *constant negative slope*. In Figure 106, I have drawn a $v_y - t$ graph with a constant negative slope. The graph also reflects the fact that at time $t = 0$ the velocity vector points slightly upward, meaning that v_y at time $t = 0$ is positive.

Notice how the $v_y - t$ graph qualitatively reflects the behavior of the velocity vector **v**: At first the velocity vector points somewhat upward; over time, the vector rotates downward until, at the peak, the vector is purely horizontal. At that moment $v_y = 0$, and the $v_y - t$ curve crosses the t-axis. From there on, the velocity vector continues rotating downward, as shown by the fact that the $v_y - t$ graph takes on values that are more and more negative.

Finally, in Figure 107 I have drawn the graph for the speed v. Remember that in general the speed v is found from the velocity components v_x and v_y by the Pythagorean Theorem: $v = \sqrt{v_x^2 + v_y^2}$. The horizontal speed remains constant, while the vertical component starts fast, drops to zero, then becomes fast again. So the speed as a whole starts fast, decreases somewhat, and then becomes fast again. The minimum speed occurs right at the top of the trajectory, when $v_y = 0$ and the speed is given entirely by the horizontal component $|v_x|$. ◇

A Concept Map for Acceleration

Figure 108 summarizes the basic ideas we have learned about acceleration.

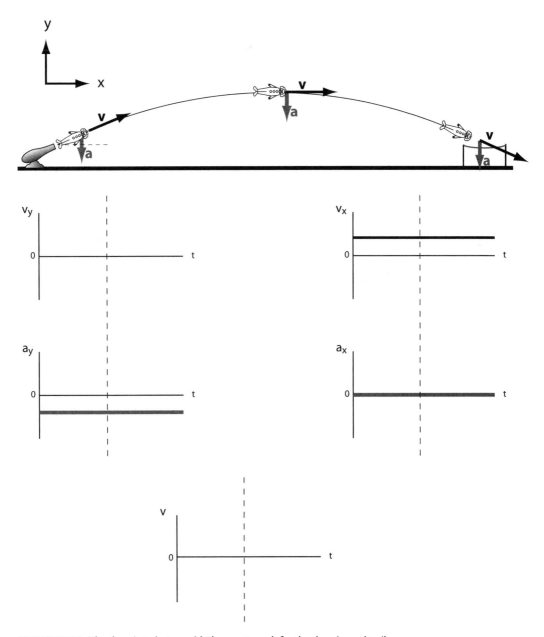

FIGURE 105. The clown's trajectory with the $a_y - t$ graph for the clown's motion (lower left-hand corner). Look back and forth between the $a_y - t$ graph and the acceleration vectors in the diagram until you can see how the graph reflects the behavior of the acceleration vector over time.

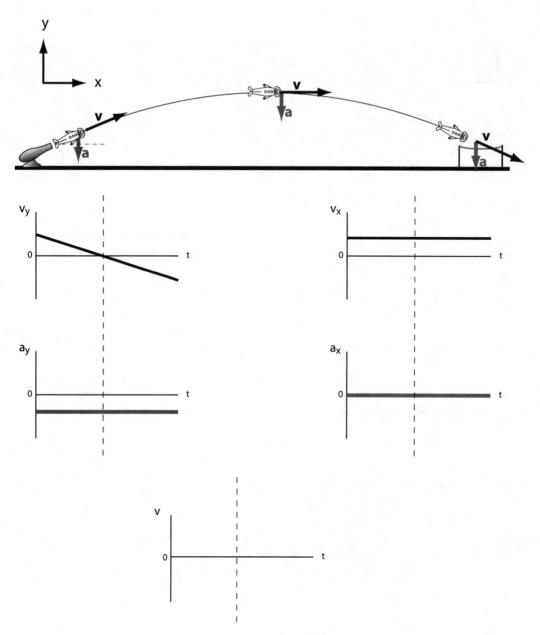

FIGURE 106. The clown's trajectory with the $v_y - t$ graph for Worked Example 3 (upper left-hand corner). Look back and forth between the $a_y - t$ graph and the $v_y - t$ graph until you can see that the $a_y - t$ graph shows the rate of change of the $v_y - t$ graph.

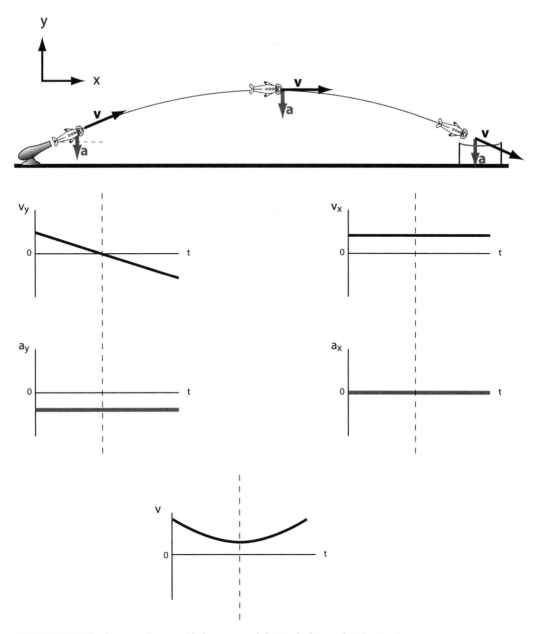

FIGURE 107. The clown's trajectory with the $v - t$ graph for Worked Example 3 (bottom). This graph combines the v_x and v_y graphs into a single graph for the overall speed of the clown.

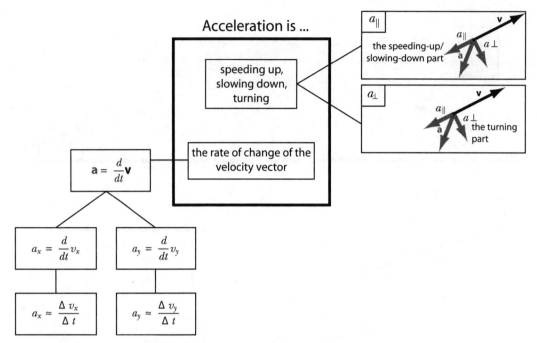

FIGURE 108. A high-level "concept map" for acceleration.

Just for the fun of it, you might try your hand at constructing a similar concept map for velocity.

Reader's Notes

Below is space for you to summarize key points from this chapter.

FOCUSED PROBLEMS :: CHAPTER 7

1. The setting for this problem is my living room floor, where at the moment a mouse is scurrying like mad to get away from my cat Spooky.

   Let's agree on a coordinate system. The x-axis passes through the center of the floor along an east-west line, and the y-axis passes through the center of the floor along a north-south line. The positive x-direction is east. The positive y-direction is north.

   At a particular instant of time, the mouse's velocity vector is $\mathbf{v} = (+3\,\text{m/s}, -3\,\text{m/s})$ and its acceleration vector is $\mathbf{a} = (50\,\text{m/s}^2, 10\,\text{m/s}^2)$.

   a. Sketch the x- and y-axes, showing the origin and the positive directions. Then sketch the mouse's velocity vector and acceleration vector.

   b. How fast is the mouse going? In what direction is it moving?

   c. Find the magnitude and direction of the mouse's acceleration vector.

   d. At this instant of time, is the mouse speeding up? Slowing down? Turning?

   e. A tenth of a second from now, approximately what will the mouse's speed be? (Hint: $\Delta v_x \approx a_x \Delta t$ and $\Delta v_y \approx a_y \Delta t$.)

2. At 3:00 p.m., an ocean liner was sighted 240 miles west of San Francisco and 120 miles north of San Francisco. At 3:15 p.m., the ocean liner was sighted 242 miles west of San Francisco and 119 miles north of San Francisco. At 3:30 p.m., the ocean liner was sighted 246 miles west of San Francisco and 117 miles north of San Francisco.

   a. Estimate the ocean liner's acceleration vector during this time period, taking the positive x-direction to be east and the positive y-direction to be north. (Take the origin of the coordinates to be San Francisco.) Give the acceleration vector in component form as $\mathbf{a} = (a_x, a_y)$, and also give the magnitude and direction of the acceleration vector.

b. Was the ocean liner speeding up during the time frame of the problem? Slowing down? Turning?

3. This problem is similar to Worked Example 3 in this chapter. Figure 109 shows several velocity and acceleration vectors for a flying saucer orbiting the earth in a circular path at a constant speed. (This situation was analyzed in Worked Example 2 of Chapter 6.) Sketch graphs showing how the quantities a_x, a_y, v_x, v_y, and v vary with time.

4. This problem is similar to Worked Example 3 in this chapter. Figure 110 shows a skateboarder traversing a half-pipe from left to right. Assume that the skateboarder is speeding up whenever she is moving downward and slowing down whenever she is moving upward. Assume that she starts from rest at time $t = 0$ and comes to rest again at the last instant shown. Sketch velocity and acceleration vectors for each of the moments shown in the figure. Then sketch graphs showing how the quantities a_x, a_y, v_x, v_y, and v vary with time.

5. [Uses calculus.] An object has a velocity vector that varies with time as $\mathbf{v}(t)$. If we assume that the object's acceleration vector $\mathbf{a}$ is nonzero at a particular time t_0, a short time later at time $t_0 + \Delta t$, the new velocity vector $\mathbf{v}(t_0 + \Delta t)$ will be given by Taylor's Theorem as

$$\mathbf{v}(t_0 + \Delta t) \approx \mathbf{v}(t_0) + \left(\frac{d}{dt} \big|_{t=t_0} \mathbf{v} \right) \Delta t \qquad (81)$$

or, in other words,

$$\mathbf{v}(t_0 + \Delta t) \approx \mathbf{v}(t_0) + \mathbf{a}(t_0) \Delta t. \qquad (82)$$

a. Assuming that $\mathbf{v}(t_0) \neq \mathbf{0}$, draw the vectors $\mathbf{v}(t_0)$, $\mathbf{v}(t_0 + \Delta t)$, and $\mathbf{a}(t_0) \Delta t$ so that they satisfy the relationship in Equation 82. Compare the results with Figure 97(b).

b. What does the picture look like if the object happens to be at rest at time t_0 (i.e., $\mathbf{v}(t_0) = \mathbf{0}$)? How does this relate to Figure 96 in Chapter 6?

SOLUTIONS CAN BE FOUND ON PAGE 366

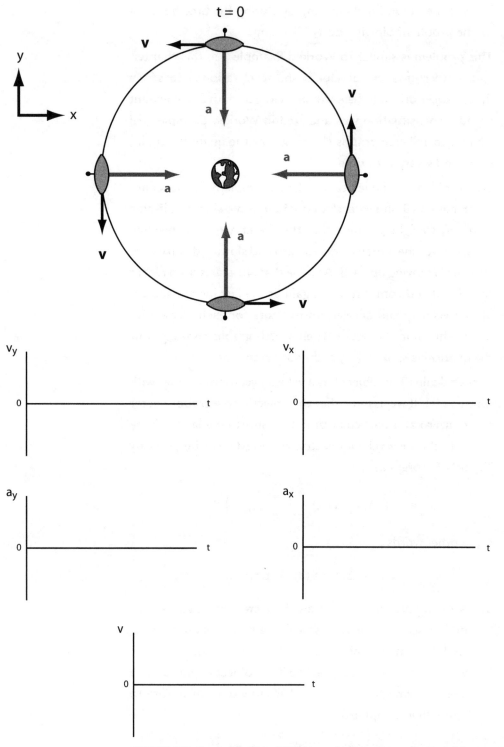

FIGURE 109. Focused Problem 3. Velocity and acceleration vectors for a flying saucer.

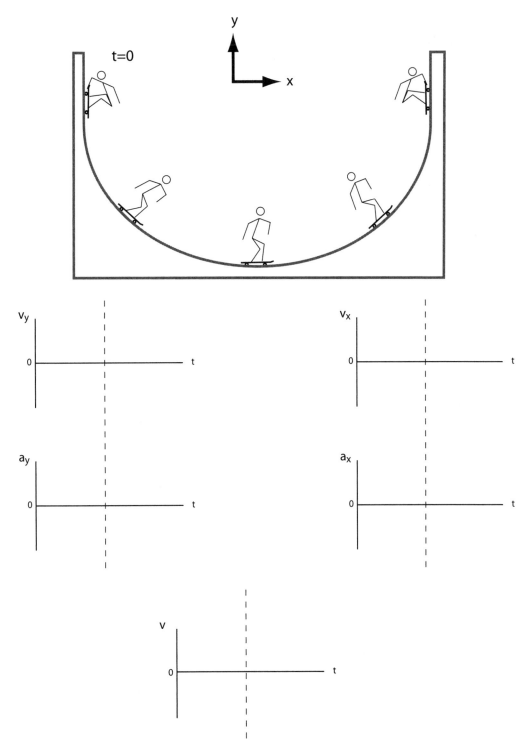

FIGURE 110. Focused Problem 4. A skateboarder skating a half-pipe.

CHAPTER 8

Focus on a-Perp

As the concept map in Chapter 7 (Figure 108) illustrates, there are two basic ways to view the acceleration vector.

- As we discussed in Chapter 6, we can view the acceleration vector as an indication of whether an object is speeding up, slowing down, or turning.
- Equally well, as we discussed in Chapter 7, we can look at the acceleration vector as an indication of how fast the velocity vector itself is changing with time.

At the intersection of these two basic perspectives lies a marvelous formula for $a_\perp$, the turning part of the acceleration. This is a formula you will use all the time in your study of physics.

The formula first appeared in 1658, in a book called *Horologium Oscillatorium*, written by Christian Huygens, a Dutch physicist. Huygens considered a very specific kind of motion: namely, an object moving in a circular path at a constant speed. This pattern of motion is called "uniform circular motion." In uniform circular motion, the speed is constant, so $a_\parallel$ is zero. The only acceleration present is the turning part, $a_\perp$. Huygens derived the following formula for $a_\perp$:

$$a_\perp = \frac{v^2}{r}.$$

(83)

Here, v is the speed of the object as it moves around the circle, and r is the radius of the circular path.

Notice that $a_\perp$ has the correct units for acceleration:

$$\frac{(m/s)^2}{m} = \frac{m^2}{s^2 \cdot m} = \frac{m}{s^2}.$$

Where Huygens's Formula Comes From

The basic insight behind Huygens's Formula is that—all other things being equal—if you move in a very tight circle (a circle with a small value of r), your velocity vector will be changing at a rapid rate. The reason for this is that when you move in a small circle, your velocity vector rotates through large angles in short times. Compare the small circular track in Figure 111 with the larger one.

Figure 111 shows snapshots of two cars moving at equal speeds but driving on circular tracks of very different size. The snapshots are shown at equally spaced moments of time t_1, t_2, and t_3. The velocity vectors are shown at these moments. For the car on the small track, the velocity vector changes a lot in the given time interval, meaning that the rate of change of velocity (acceleration) is large. For the car on the large track, the velocity vector changes much less in the given time interval, meaning that the rate of change of velocity (acceleration) is small. Therefore, the acceleration will be greater when the radius is smaller. This is why there is an r in the denominator of Huygens's Formula.

Of course, the radius of the circle isn't everything. Your speed matters, too, because the faster you move around the circle, the faster your velocity vector will be rotating through those angles. This is why there is a v^2 in the numerator of the formula: A large value of v^2 means that you are traveling the circle quickly, so your velocity vector is changing fast; hence we should have a large acceleration.

If you want to know more of the details leading to Huygens's Formula, the Appendix gives a mathematical derivation.

Beyond Uniform Circular Motion

Huygens derived Equation 83 for the particular case of motion in a circular path at a constant speed. But the wonderful thing about Huygens's Formula is that it works even when the speed is

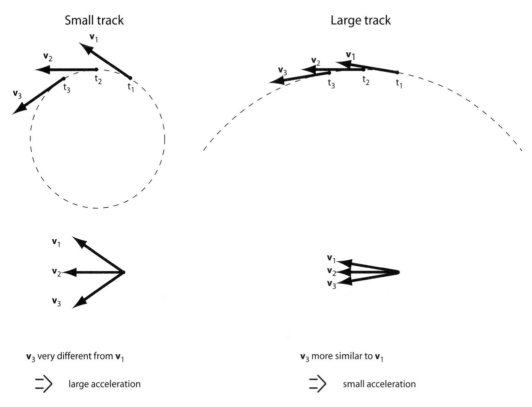

FIGURE 111.
Snapshots of two
cars moving at equal
speeds on circular
tracks of different
size.

not constant and even when the trajectory is *not* circular! (I don't know if Huygens himself knew this or not.)

You can use $a_\perp = \frac{v^2}{r}$ to find the perpendicular component of the acceleration vector pretty much no matter *how* an object is moving. You just have to know how to interpret the "r" in the formula when the path is not a circle.

In general, the r in Huygens's Formula is the *radius of curvature* of the object's trajectory. See Figure 112.

When you're moving in a curved path, the radius of curvature is an indication of how tightly you are turning at any given time. Tight curves have a small radius of curvature, while gentle curves have a large radius of curvature. A straight trajectory (not curved at all) has an *infinite* radius of curvature ($r \to \infty$). Intuitively, this is because a straight path is like a portion of an infinitely large circle.

Computing the radius of curvature of a given trajectory at a given point is a calculus problem. You don't need to know how to do this; I'll be giving

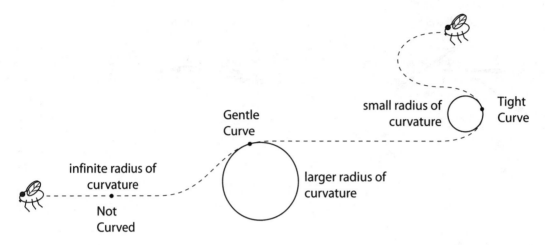

Gentle
Curve

small radius of
curvature

Tight
Curve

infinite radius of
curvature

larger radius of
curvature

Not
Curved

FIGURE 112.
The radius of
curvature at three
different points of
a fly's trajectory.

you the radius of curvature in any problems we have to solve. Just develop
your intuitive sense that tight curves have a small radius of curvature,
while gentle curves have a large radius of curvature.

In Chapter 6 you sketched velocity and acceleration vectors for
a person walking at a constant speed along an S-shaped path (see
Worked Example 5 and Figure 94 in Chapter 6). In that problem,
a was perpendicular to **v** because the person walks at a constant
speed (not speeding up or slowing down, only turning). Back then,
we weren't concerned about the *magnitude* of the acceleration vector, only about the direction. But with Huygens's Formula in hand,
we can improve our sketches a bit.

» WORKED EXAMPLE 1

Intuitively speaking, how will the magnitude of the acceleration
vector vary as the person traces out the S-shaped path at constant
speed?

» SOLUTION 1

The speed is constant, so $a_\parallel = 0$. This means that the acceleration
vector **a** is entirely perpendicular to **v**. In this case, the accelera-
tion vector **a** has magnitude $a = \sqrt{a_\parallel^2 + a_\perp^2} = \sqrt{0 + a_\perp^2} = a_\perp = \frac{v^2}{r}$.
Therefore, in our sketches we should make the acceleration vector
long where the curve is tight, and we should make the acceleration
vector short where the curve is gentle. See Figure 113. ◇

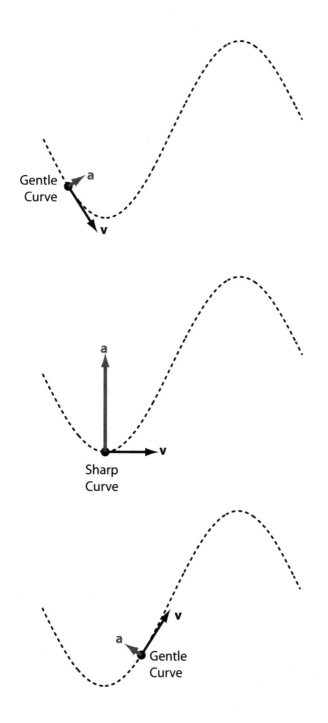

FIGURE 113. Worked Example 1. Acceleration vectors for gentle and sharp curves.

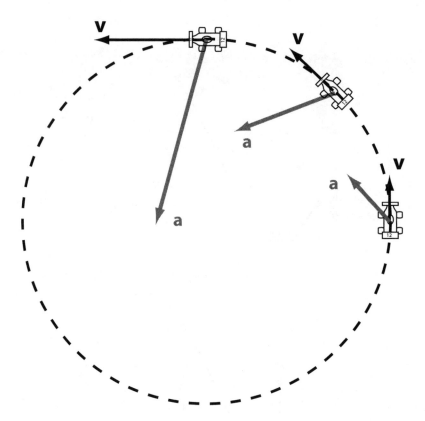

FIGURE 114.
Solution to Worked Example 2. My sketches of a race car's velocity and acceleration vectors at three moments in time.

»WORKED EXAMPLE 2

A toy race car is going around a circular track. The speed of the race car is continually increasing. Sketch the race car's velocity and acceleration vectors at three equally spaced moments in time.

»SOLUTION 2

My sketches are shown in Figure 114. Notice that the velocity vector is lengthening over time.

I drew the acceleration vectors partly along the velocity vector (because the race car is speeding up) and partly perpendicular to the race car (because the car is turning). Because v is becoming larger while r stays the same, we know that $a_\perp = \frac{v^2}{r}$ is becoming larger, so I drew my acceleration vector with the perpendicular component becoming larger over time. $\diamond$

Huygens's Formula is one more piece of information we can add to our concept map for acceleration; see Figure 115.

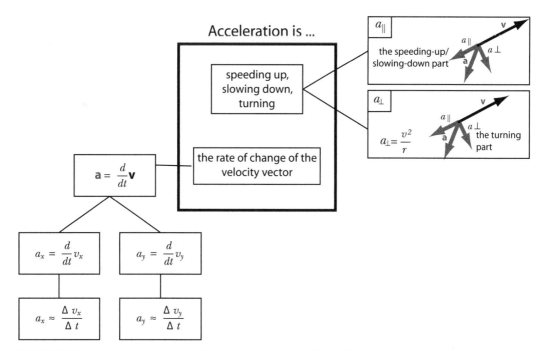

FIGURE 115. A high-level concept map for acceleration, now including Huygens's Formula, $a_{\perp} = \frac{v^2}{r}$.

Reader's Notes

Below is space for you to summarize key points from this chapter.

FOCUSED PROBLEMS :: CHAPTER 8

1. When you roll a penny on its edge along a tabletop, the penny will sometimes perform a spiraling motion. The penny spirals toward the center before it falls. Assume that a penny is rolling in such a spiral pattern at a constant speed. Sketch the penny's velocity and acceleration vectors at three different moments in time.

2. Figure 116 shows a top-down view of a car driving on cruise control at a constant speed on a curved road. Sketch the car's velocity and acceleration vectors at the indicated moments of time.

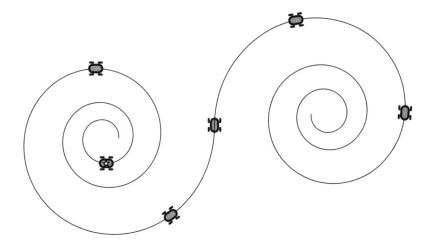

FIGURE 116. Focused Problem 2. Snapshots of a car driving on a curved road. The snapshot marked with the star is the first snapshot in the series.

3. Assume that the earth orbits around the sun at a constant speed in a circular path of radius $r_0 = 1.5 \times 10^{11}$ m. (This is not far from the truth.)

    a. Sketch the earth's orbit with the sun at the center, showing the earth's velocity and acceleration vectors at four equally spaced moments throughout the year.

    b. Calculate the earth's speed, in meters per second, as it revolves around the sun. (Hint: For this just use the grade-school formula "speed = distance/time.")

    c. Calculate the magnitude of the earth's acceleration vector. Label your sketch with this information.

4. A car pulls out of a driveway onto the main road, speeding up as it turns. See Figure 117.

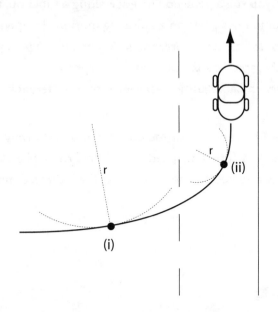

At point (i), the car's speed is 5 m/s and the radius of curvature of its path is 6 m. At point (ii), the car's speed is 8 m/s and the radius of curvature of its path is 2 m.

a. Sketch the car's velocity and acceleration vectors at both points.

b. Compute the numerical value of $a_\perp$ at the two instants of time.

c. If the car is speeding up at a constant rate of 3 m/s²—this is $a_\parallel$, by the way—compute the magnitude of the (total) acceleration vector at the two instants of time.

5. A car is driving on cruise control at a constant speed v_0. The road is extremely hilly, as shown in Figure 118.

The car and its velocity vector are shown at seven different instants of time along its path. On the figure, sketch the car's acceleration vector at each of these instants. If the acceleration vector is zero at any of these instants of time, just write "$a = 0$" there.

Make the acceleration vectors to scale, in the sense that if one acceleration is larger in magnitude than another, the

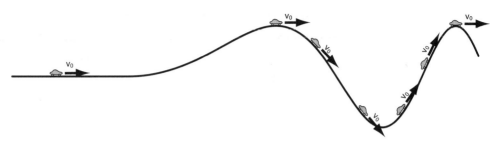

FIGURE 118. Focused Problem 5. Snapshots of a car driving at a constant speed on a hilly road.

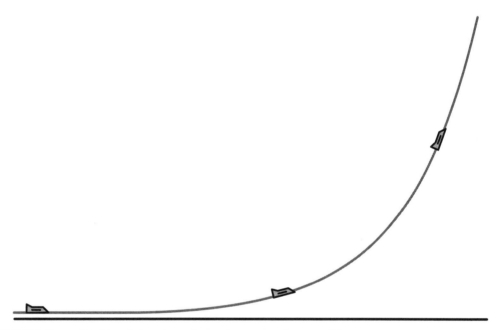

FIGURE 119. Focused Problem 6. Snapshots of a jet plane at three instants of time during takeoff.

arrow is noticeably larger, while if two accelerations are equal, the arrows are about equally long. Also, be careful to draw the acceleration vectors so that they point in the direction in which you mean them to point.

6. Figure 119 shows a jet plane at three instants of time during a radical takeoff maneuver. Assume that the plane was already in motion when the first snapshot was taken. Also, assume that the plane was continually speeding up during the entire maneuver (in all three snapshots).

On the figure, sketch the jet's velocity vector and acceleration vector at each of the three instants of time. If the acceleration vector is zero at any of the three instants of time, just write "$a = 0$" there.

SOLUTIONS CAN BE FOUND ON PAGE 367

CHAPTER 9

Case Study: Straight-Line Motion

The richness of Newton's Laws can be appreciated only in the context of curvilinear motion. The classic touchstone for full understanding is the simple pendulum, with its dynamic interplay of tension, gravity, acceleration, and velocity vectors. From this perspective, straight-line motion is a time-consuming topic that does very little to prepare the mind for what is to come. At the same time, however, the very fact that physics is a vectorial subject means that the world *can* be analyzed one component at a time. So the tools introduced in this chapter turn out to be crucial! The point is, if you want to take advantage of what we've already discussed about acceleration and find out how it relates to force, skip this chapter and go on—you'll be pursuing the main thread of ideas if you do. But at some point, you'll need to come back.

It would be a very good idea to reread Chapter 2 in conjunction with Chapter 9.

Sometimes there are simple situations in which the motion of an object is confined more or less to a straight line. Think of driving a car along a straight (and flat) road, stepping off of a diving board and falling straight down, working a yo-yo straight up and down, or riding on an escalator.

Looking at Vectors r, v, and a in Straight-Line Motion

To look in depth at an example of straight-line motion, let's think about the motion of a trolley car on a straight horizontal track. The trolley car can start, stop, speed up, and slow down. See Figure 120.

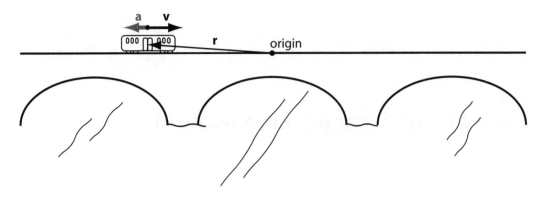

FIGURE 120.
Motion of a trolley
car on a straight
horizontal track.

At the instant of time shown, its position vector is **r**, its velocity vector is **v**, and its acceleration vector is **a**.

To look at these vectors quantitatively, we have to lay down a coordinate system. In Figure 121, I have laid down a coordinate system with the origin at the center of the track. The x-axis falls along the track, with x-values increasing to the right, and the y-axis falls perpendicular to the track, with y-values increasing upward.

With a coordinate system, we can talk about the components of the position, velocity, and acceleration vectors.

Position vector. The position vector **r** will always point from the origin directly to the center of the car. (This is the point of the car that we will consider as defining its overall location.) Because the trolley car is constrained to move along the track, the y-component of the position vector will always have the value $r_y = +2$ m or so.

Meanwhile, the x-component of the position vector will vary with time. So we will have $r_x = r_x(t)$ or, in abbreviated terms, $x = x(t)$.

Velocity vector. The velocity vector **v** will always have a y-component of zero: $v_y = 0$. One way to see this is to recognize that the velocity vector of the trolley car always points either purely to the right or purely to the left, so it never has a component along the y-axis.

Alternatively, there is another way to see why $v_y = 0$. Remember that v_y is the rate of change of y; because the value of y remains constant over time, its rate of change is zero.

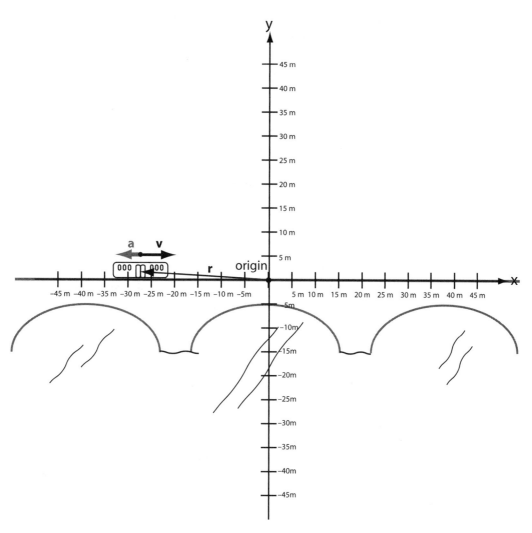

FIGURE 121.
A coordinate system overlaid on Figure 120.

Because the velocity vector has a y-component of zero, the speed of the trolley car is given by the absolute value of v_x. This is because $v = \sqrt{v_x^2 + v_y^2} = \sqrt{v_x^2 + 0} = \sqrt{v_x^2} = |v_x|$.

Acceleration vector. The acceleration vector **a** will also maintain a y-component of zero: $a_y = 0$. One way to see this is to recognize that although the trolley car may be speeding up or slowing down, it is never turning. This means that we would always sketch the acceleration vector pointing either directly along the velocity vector or directly against the velocity vector. Either way, the acceleration vector points purely in the positive or the negative x-direction and has no y-component.

Another way to see why $a_y = 0$ is to remember that a_y is the rate of change of v_y; because the value of v_y remains constant over time, its rate of change is zero.

Because the acceleration vector has a y-component of zero, the magnitude of the acceleration vector is given by the absolute value of a_x. This is because $a = \sqrt{a_x^2 + a_y^2} = \sqrt{a_x^2 + 0} = \sqrt{a_x^2} = |a_x|$.

» WORKED EXAMPLE 1

For each trolley car shown in Figure 122, what is the horizontal position x of the center of the car? What is the vertical position y of the center of the car? For the rightmost car, say whether the x- and y-components of the position, velocity, and acceleration vectors are positive, negative, or zero. How would you describe the motion of this car at the instant of time shown?

» SOLUTION 1

The center of the leftmost car has a horizontal position of $x_1 = -35$ m and a vertical position of $y_1 = +2$ m or so. The second car has a horizontal position of $x_2 = -12.5$ m and a vertical position of $y_2 = +2$ m. The third car has a horizontal position of $x_3 = +20$ m and a vertical position of $y_3 = +2$ m. The rightmost car has a horizontal position of $x_4 = +35$ m and a vertical position of $y_4 = +2$ m.

Focusing now on the rightmost car, the x-component of $\mathbf{r}$ is positive ($+35$ m); the y-component of $\mathbf{r}$ is positive ($+2$ m); the x-component of $\mathbf{v}$ is negative; the y-component of $\mathbf{v}$ is zero; the x-component of $\mathbf{a}$ is positive; and the y-component of $\mathbf{a}$ is zero.

Notice that $\mathbf{a}$ points against $\mathbf{v}$, indicating that the car is slowing down at the instant of time shown. All in all, at the instant of time shown, the car is located to the right of the origin, moving to the left, and in the process of slowing down. ◇

Position as a Function of Time: $x - t$ Graphs

If the trolley car moves back and forth from one end of the track to the other, its position will vary with time. This means that the number x that describes its position will also vary with time. So the motion of the trolley car can be described by a function $x(t)$ that specifies the position x at any given time t. Graphing this

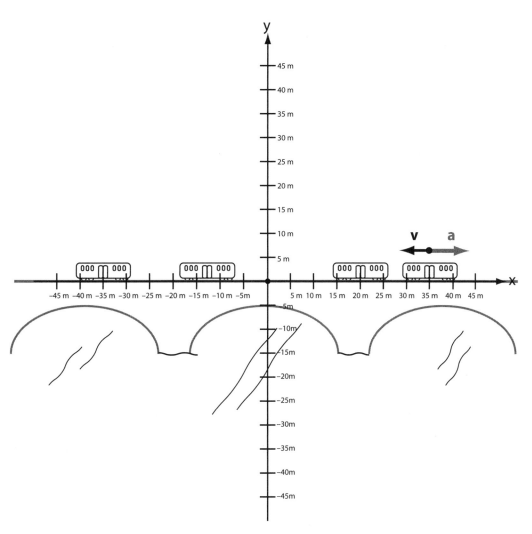

FIGURE 122.
Worked Example 1.
Determining the
position, velocity,
and acceleration
vectors of four trolley
cars and describing
their motion.

function is a way to gain insight into the motion of the car. We call such graphs position-time graphs, or $x - t$ graphs for short. See Figure 123.

Careful! The actual motion of the car is back and forth, from left to right. But in an $x - t$ graph the position axis is usually drawn vertically. The mathematical curve varies in an up-and-down fashion, but when we translate this back into our cartoon, it means the car is moving left and right. I find it helps to view the vertical axis of an $x - t$ graph as a "copy" of your number-line coordinate system stood up on its end.

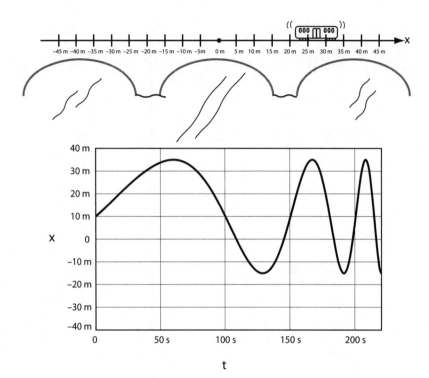

FIGURE 123.
A trolley car moving
back and forth and
the position-time
$(x - t)$ graph of its
motion.

There is a lot of information hidden in position-time graphs. Here is an example of what you can see when you study them closely.

» WORKED EXAMPLE 2

Based on the $x - t$ graph in Figure 123, during which times is the car located to the right of the origin? During which times is the car located to the left of the origin? When is the car farthest from the origin? During which times is the car moving to the right? During which times is the car moving to the left? During which times is the car moving fastest? During which times is the car at rest (stopped)? Indicate all of these times on the figure.

» SOLUTION 2

My answers are summarized in Figure 124. Note that five of the vertical lines are dashed, indicating that the trolley car is passing through the origin, while the other five are dotted, indicating that the car is at rest.

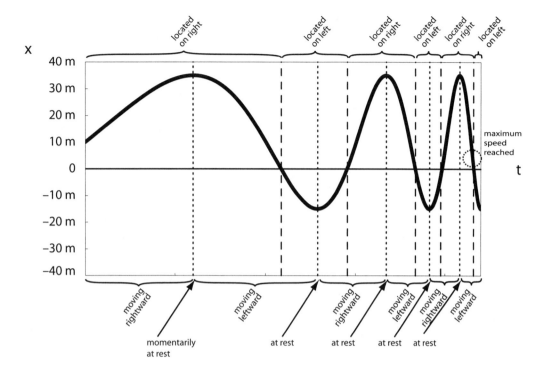

FIGURE 124.
My answers to Worked Example 2, about the trolley car in Figure 123.

Let's see how I answered this question. To begin with, you always have to refer to a coordinate system to answer questions about motion. So look at the coordinate system in Figure 123. According to the coordinate system, positive numbers are used to refer to locations on the right of the origin. Negative numbers are used to refer to locations on the left of the origin. Looking at the $x - t$ graph, the trolley car starts off at $+10$ m, which is 10 m to the right of the origin. Periods when the car is located to the right of the origin are indicated along the top of the graph.

When the $x - t$ curve reaches $x = 0$, that's when the trolley car passes through the origin. This happens several times during the motion sequence, as indicated on the $x - t$ graph by vertical dashed lines.

The trolley car is farthest from the origin at the times indicated by the first, third, and fifth dotted lines; here it is about 35 m away from the origin.

The coordinate system also tells us that motion to the right corresponds to motion toward *increasing x values*. If the trolley car moves to the right along the number line in Figure 123, its x-values become larger.

On the $x - t$ graph in Figure 124, the trolley car's x-value is increasing from the starting time until it reaches the first vertical dotted line. The trolley car is always moving to the right during this period. All in all, there are three different periods during which the trolley car is moving rightward. These periods are indicated along the bottom of the graph.

The coordinate system also tells us that motion to the left corresponds to motion toward *decreasing x values*. If the trolley car moves to the left along the number line in Figure 123, its x-values become smaller.

On the $x - t$ graph in Figure 124, the x-value is decreasing from the first vertical dotted line to the next vertical dotted line. The trolley car is always moving to the left during this period. All in all, there are three different periods of time during which the trolley car is moving leftward. These periods are indicated along the bottom of the graph.

Now for the speed of the car's motion. The speed is given by $|v_x|$, the absolute value of the x-velocity. And v_x is $\frac{d}{dt}x$, the rate of change of the x-position. So in visual terms, v_x is given by the *slope* of the $x - t$ graph.

The steepest point of the $x - t$ graph occurs in the vicinity of the dotted circle. This is where the car is going fastest. (Here $\frac{d}{dt}x$ is negative, so the car is moving leftward at this moment.)

The car is at rest when its position is not changing—that is, when its rate of change of position is zero. To find these times, look for places on the $x - t$ graph where the slope is zero, that is, where the curve is horizontal. This happens at times indicated by arrows. These are all times when the car is (momentarily) at rest.

After you have analyzed the $x - t$ graph used to describe the trolley car's motion, try to reflect on what you have learned until you have a sense of what the overall motion is like. Try to tell a "story" about the motion, or even act it out yourself by walking back and forth.

What would the motion of our trolley car look like? In this case, the story would be about an out-of-control trolley car madly cycling back and forth between stations at an ever-increasing speed! Not very pleasant for the passengers. ◇

Velocity as a Function of Time: v_x – t Graphs

An $x - t$ graph shows the x-component of the position of a moving object as a function of time. It is also useful to analyze $v_x - t$ graphs, which show the x-component of the velocity of a moving object as a function of time.

» WORKED EXAMPLE 3

At time $t = 0$, a dragster starts from rest and gradually speeds up at a constant rate, reaching a top speed of 300 mph at time $t = 6$ s. The dragster holds this speed until it crosses the finish line at $t = 12$ s. As it crosses the finish line, a parachute at the back opens and the dragster slows down at a constant rate, coming to rest at time $t = 15$ s. The dragster sits still for 30 seconds and is then towed back to the starting line at a constant speed of 5 mph. (i) Draw a cartoon of the drag strip, and lay down a coordinate system. (ii) Draw five cartoons that show the dragster during the five different phases of the motion: the speeding-up phase, the coasting phase, the braking phase, the resting phase, and the return phase. On each cartoon, draw the position, velocity, and acceleration vectors for the dragster. (iii) Finally, draw a numerically accurate $v_x - t$ graph for the entire motion.

» SOLUTION 3

My coordinate system, vectors, and $v_x - t$ graph are shown in Figure 125. As a "bonus" I have also sketched **a** and **v** for the very initial moment at $t = 0$; at this moment, the dragster is not yet moving, but it is in the process of speeding up. (See Figure 96 in Chapter 6 for a similar situation.)

I sketched my velocity vectors so that their relative lengths indicate the relative speeds of the dragster at the different times indicated. Faster motion means the velocity vector is longer.

I sketched my acceleration vectors so that they indicate speeding up (**a** along **v**) or slowing down (**a** against **v**), as appropriate. I also made the relative lengths of the acceleration vectors indicative of the rate of speeding up or slowing down. The more sudden the change, the longer I made the acceleration vector.

As the $v_x - t$ graph shows, there is a brief speeding-up phase beginning at $t = 45$ s; this is when the dragster begins being towed backward, and its speed has to build up from $v = 0$ to $v = 5$ mph. The problem doesn't tell us how long this takes; I just made it a brief interval. The acceleration vector during this brief interval is not pictured, but it points to the left during this time. ◇

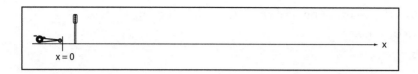

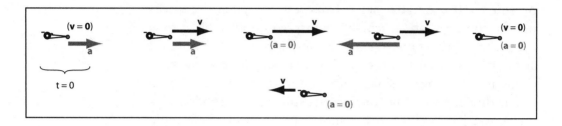

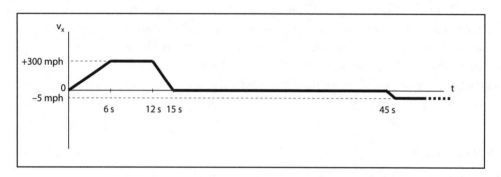

FIGURE 125.
One solution to
Worked Example 3,
about a dragster
starting from rest
and gradually
speeding up at a
constant rate.

A different possible answer to this question is shown in Figure 126, which uses a different coordinate system. Notice how the vectors in both answers are the same. However, the $v_x - t$ graph does change when you change the coordinate system because the components of a vector are always defined relative to the coordinate system you lay down.

Acceleration as a Function of Time: $a_x - t$ Graphs

An $x - t$ graph shows the x-component of the position of a moving object as a function of time. A $v_x - t$ graph shows the x-component of the velocity of a moving object as a function of time. It is also possible to make an $a_x - t$ graph, which shows the x-component of the acceleration of a moving object as a function of time.

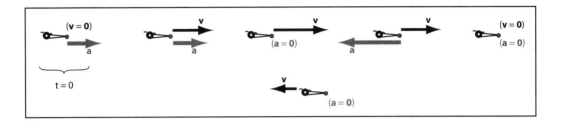

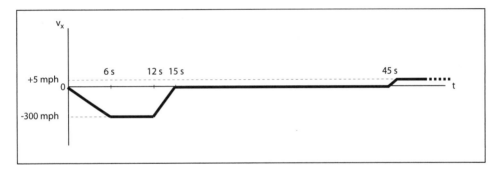

FIGURE 126.
An alternative solution to Worked Example 3.

» WORKED EXAMPLE 4

For the dragster in Worked Example 3, use the cartoons you already
drew to sketch a rough $a_x - t$ graph for the entire motion.

» SOLUTION 4

My $a_x - t$ graph is shown in Figure 127, based on the coordinate
system in Figure 125.

I made a_x positive where the acceleration vector points *with* the
arrowhead on the x-axis, and I made a_y negative where the acceleration vector points *against* the arrowhead on the x-axis. And I made
the absolute value $|a_x|$ reflect the magnitude of the acceleration vec-
tor. The $a_x - t$ graph would have looked different if I had followed
the coordinate system shown in Figure 126. But all of the vectors
are the same in both figures. ◇

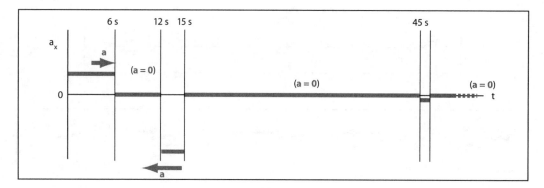

FIGURE 127.
Solution to Worked
Example 4. My $a_x - t$
graph based on the
coordinate system
for the dragster's
motion shown in
Figure 125.

Sketching v_x – t Graphs from x – t Graphs

If you are given an $x - t$ graph that describes the motion of an object, it is easy to sketch a $v_x - t$ graph for the motion of the object. Because velocity is the rate of change of position, all you have to do is sketch the rate of change.

The $v_x - t$ graph is the slope of the $x - t$ graph.

» WORKED EXAMPLE 5

For the $x - t$ graph of the trolley car shown in Figure 123, sketch the corresponding $v_x - t$ graph illustrating the trolley car's velocity as a function of time.

» SOLUTION 5

Figure 128 shows my sketch of the rate-of-change graph. I always like to line up my graphs so that the $x - t$ graph is on top and the $v_x - t$ graph is right below that. This way, I can look from one graph to the other as I sketch the slope.

Did you start your $v_x - t$ graph off with a positive value, as I did? Look at the $x - t$ graph: the slope is positive right at the start.

Notice the five dashed lines; these illustrate times when the trolley car is at rest. What is special about these times on the $x - t$ curve? What is special about these times on the $v_x - t$ curve?

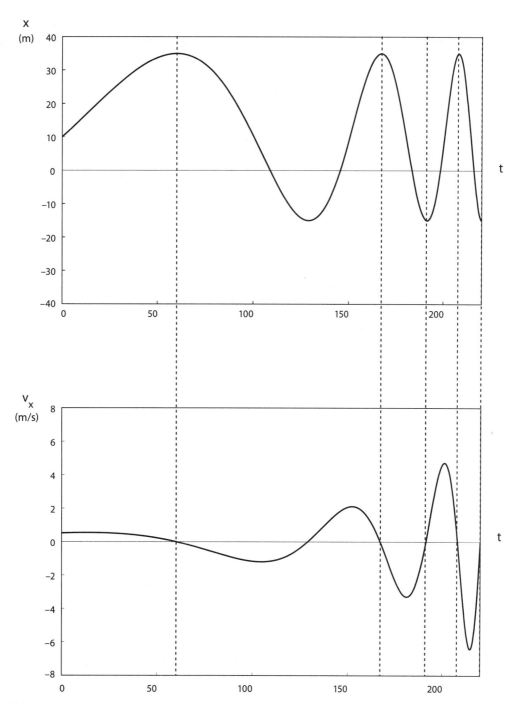

FIGURE 128. Solution to Worked Example 5. My $v_x - t$ graph for the trolley car shown in Figure 123.

Notice, too, how the $v_x - t$ graph suggests the general speeding up of the trolley car. The peak speeds toward the end of the motion are higher than the peak speeds toward the beginning. ◇

Sketching $a_x - t$ Graphs from $v_x - t$ Graphs

If you are given a $v_x - t$ graph that describes the motion of an object, it is easy to sketch an $a_x - t$ graph for the motion of the object. Because acceleration is the rate of change of velocity, all you have to do is sketch the rate of change.

The $a_x - t$ graph is the slope of the $v_x - t$ graph.

»WORKED EXAMPLE 6

For the dragster in Worked Examples 3 and 4, the $v_x - t$ graph is shown in Figure 125 and the $a_x - t$ graph is shown in Figure 127. For convenience, the two graphs are shown aligned in Figure 129.

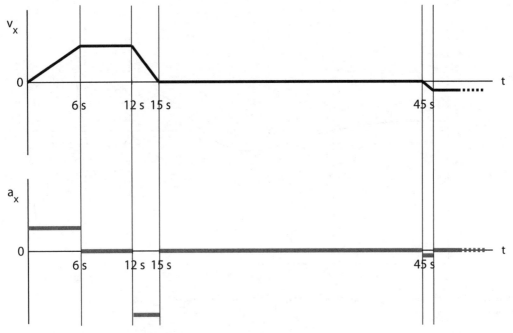

FIGURE 129. Worked Example 6. The $v_x - t$ graph for the dragster shown in Figure 125 aligned with the $a_x - t$ graph shown in Figure 127.

By looking at the two graphs, verify that, in qualitative terms, the *value* of the $a_x - t$ graph does indeed describe the *slope* of the $v_x - t$ graph at each point.

» SOLUTION 6

From $t = 0$ to $t = 6$ s, the slope of the v_x graph is positive, and during this interval the value of a_x is indeed positive. From $t = 6$ s to $t = 12$ s, the slope of the v_x graph is zero, and during this interval the value of a_x is indeed zero. From $t = 12$ s to $t = 15$ s, the slope of the v_x graph is negative (and steep), and during this interval the value of a_x is indeed negative (and large in absolute value). From $t = 15$ s to $t = 45$ s, the slope of the v_x graph is zero, and during this interval the value of a_x is indeed zero. Next comes the brief speeding up as the tow truck brings the dragster up to speed; here the slope of the $v_x - t$ curve is negative, and the value of the $a_x - t$ curve is indeed negative. After that, the slope of the v_x graph is zero, and during this interval the value of a_x is indeed zero. ◇

Some Helpful Observations about $x - t$, $v_x - t$, and $a_x - t$ Graphs

All of the information in an object's $v_x - t$ and $a_x - t$ graphs is already contained in the object's $x - t$ graph. But the same information is represented differently in the various graphs. For example, suppose an object is moving *with the arrow* of the coordinate system. On the $x - t$ graph this fact will be represented by an *uphill slope*. But on the $v_x - t$ graph this fact will be represented by a *positive value*.

Now for another example. Suppose the object is *momentarily at rest.* On the $x - t$ graph this fact will be represented by a *momentarily horizontal slope.* But on the $v_x - t$ graph this fact will be represented by the curve *touching the t-axis.*

Here's one more example. Suppose the object is *moving against the x-arrowhead and speeding up.* On the $v_x - t$ curve this fact will be represented by *a negative value and a negative slope.* On the $x - t$ curve this fact will be represented by a *downhill slope and a downward concavity.*

These observations and more are summarized in Table 3.

Notice that the acceleration at $t = 0$ is *not* zero! This is one of those tricky cases in which an object is accelerating but not (yet) moving. You might want to go back and reread "A Tricky Case" in Chapter 6. Also, I think you must agree that in Figure 129 the slope of the $v_x - t$ graph at $t = 0$ is positive, not zero.

Do not memorize Table 3! Instead, make sure that each cell in the table makes sense to you based on what you understand about position, velocity, and acceleration.

TABLE 3. **Some Observations on** $x - t$, $v_x - t$, **and** $a_x - t$ **Graphs**

Situation	What the $x - t$ curve will look like in that region	What the $v_x - t$ curve will look like in that region	What the $a_x - t$ curve will look like in that region
Object located at origin	Curve touches t-axis	Can be anything	Can be anything
Object located on arrowhead side of origin	Curve above t-axis	Can be anything	Can be anything
Object located on other side of origin	Curve below t-axis	Can be anything	Can be anything
Object at rest	Curve horizontal	Curve touches t-axis	Can be anything
Object moving with the arrowhead	Curve has +slope	Curve above t-axis	Can be anything
Object moving against the arrowhead	Curve has −slope	Curve below t-axis	Can be anything
Object moving with the arrowhead and speeding up	Curve has +slope and is concave up	Curve above t-axis with + slope	Curve above t-axis
Object moving with the arrowhead and slowing down	Curve has +slope and is concave down	Curve above t-axis with −slope	Curve below t-axis
Object moving against the arrowhead and speeding up	Curve has −slope and is concave down	Curve below t-axis with −slope	Curve below t-axis
Object moving against the arrowhead and slowing down	Curve has −slope and is concave up	Curve below t-axis with +slope	Curve above t-axis
Object momentarily at rest, about to be moving with the arrowhead	Curve momentarily horizontal, concave up	Curve touches t-axis with +slope	Curve above t-axis
Object momentarily at rest, about to be moving against the arrowhead	Curve momentarily horizontal, concave down	Curve touches t-axis with −slope	Curve below t-axis

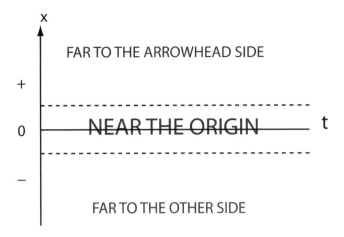

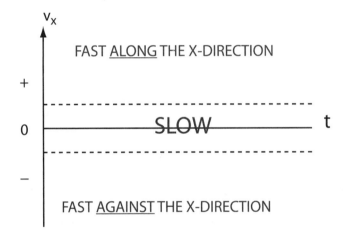

FIGURE 130.
Interpreting $x - t$
and $v_x - t$ graphs.

Figure 130 may also help you to interpret $x - t$ and $v_x - t$ graphs.

Keeping these things straight can be a real headache, and it takes practice. If you are having lots of trouble, your real difficulties may well be with the material in Chapter 2. Position-time, velocity-time, and acceleration-time graphs are just a special case of the techniques we learned in that chapter—as we shall see in more detail next.

Now would be a good time to try Focused Problems 1 through 6, which deal with the material we've covered so far in this chapter.

Net Change in Position

In Chapter 2 we learned that the net change in a quantity Q over a specified time interval is given by the area under the curve in its rate-of-change graph. Because velocity is the rate of change of position, we can find the net change in an object's x-position by finding the area under the object's $v_x - t$ graph.

To be more precise, the net change in an object's x-position over a time interval from initial time t_i to final time t_f is given by the following formula:

$$x_f - x_i = \text{net signed area bounded by the } v_x - t \text{ curve}$$
$$\text{and the } t\text{-axis, from } t_i \text{ to } t_f.$$

(84)

The term *signed area* means that we count the area as positive where the $v_x - t$ curve is above the t-axis and negative where the $v_x - t$ curve is below the t-axis.

Equation 84 generalizes the simple grade-school formula "distance = speed × time." Unlike the grade-school formula, Equation 84 works even when the speed and direction of the motion vary.

» WORKED EXAMPLE 7

Going back to the dragster in Worked Example 3, use your $v_x - t$ graph (Figure 125) to find the distance traveled by the dragster during the race.

» SOLUTION 7

The distance traveled by the dragster during the race will be the difference between its initial position at time $t = 0$ and its position at time $t = 12$ s, when it crosses the finish line. The net change in the position Δx can be found by computing the area under the $v_x - t$ curve between $t = 0$ and $t = 12$ s. See Equation 84. The area we seek is shaded in Figure 131.

The area is easily computed using the dimensions on the figure:

(85) $$\Delta x = (\text{area of triangular region}) + (\text{area of rectangular region})$$

(86) $$= \frac{1}{2}(6 \text{ s})(300 \text{ mph}) + (12 \text{ s} - 6 \text{ s})(300 \text{ mph})$$

(87) $$= 900 \text{ mph} \cdot \text{s} + 1{,}800 \text{ mph} \cdot \text{s}$$

(88) $$= 2{,}700 \text{ mph} \cdot \text{s}.$$

In normal units, this is $2{,}700\frac{\text{mi}}{\text{hr}} \times \text{s} = 2{,}700\frac{\text{mi}}{3{,}600 \text{ s}} \times \text{s} = 0.75$ mi.

◇

Net Change in Velocity

This topic is a lot like the last one. Because acceleration is the rate of change of velocity, we can find the net change in an object's x-velocity by finding the area under the object's $a_x - t$ graph.

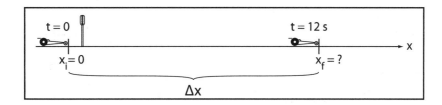

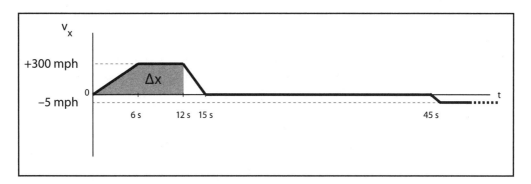

FIGURE 131.
Solution to Worked
Example 7. The
distance traveled by
a dragster during a
race.

To be more precise, the net change in an object's x-velocity over a time interval from initial time t_i to final time t_f is given by the following formula:

$$v_{f,x} - v_{i,x} = \text{net signed area bounded by the } a_x - t \text{ curve}$$
$$\text{and the } t\text{-axis, from } t_i \text{ to } t_f. \tag{89}$$

The term *signed area* means that we count the area as positive where the $a_x - t$ curve is above the t-axis and negative where the $a_x - t$ curve is below the t-axis.

» WORKED EXAMPLE 8

Going back to the dragster in Worked Example 3, calculate the x-acceleration of the dragster during the braking phase. Use area reasoning based on the known net change in velocity.

» SOLUTION 8

Based on the information given in Worked Example 3, the net change in x-velocity is $v_{x,f} - v_{x,i} = (0\,\text{mph}) - (+300\,\text{mph}) = -300\,\text{mph}$. Meanwhile, the area under the $a_x - t$ curve during the 3-second braking phase is given by height $\times$ width $= (a_x)(3\,\text{s})$, where a_x is unknown. See Figure 132.

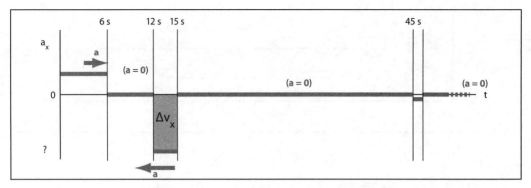

FIGURE 132. Solution to Worked Example 8. The x-acceleration of the dragster in Worked Example 3 during the braking phase.

Equating the net change in v_x to the area under the $a_x - t$ curve (in accordance with Equation 89), we have

(90) $$-300 \text{ mph} = (a_x)(3 \text{ s}),$$

which yields

(91) $$a_x = -100 \text{ mph/s}.$$

Notice the negative sign, which matches the qualitative $a_x - t$ graph in Figure 132.

The dragster loses 100 mph every second during the braking phase! This would be tough on the driver. In standard units, our answer is

(92) $$a_x = -100 \frac{\text{mph}}{\text{s}}$$

(93) $$= -100 \frac{\text{mile}}{\text{hr}} \times \frac{1}{\text{s}}$$

(94) $$= -100 \frac{\text{mile}}{\text{hr}} \times \frac{1{,}609 \text{ m}}{\text{mile}} \times \frac{\text{hr}}{3{,}600 \text{ s}} \times \frac{1}{\text{s}}$$

(95) $$= -44.7 \text{ m/s}^2.$$

This is almost "5gs" worth of acceleration! ◇

In that last problem, we could also have found a_x more directly by calculating the slope of the $v_x - t$ curve. This is perhaps the more obvious approach. My goal was to illustrate area-net change reasoning, not slope-rate-of-change reasoning. But if you did want to find a_x by slope-rate-of-change reasoning, you would get the same answer:

$$a_x = \text{slope of } v_x - t \text{ curve} \tag{96}$$

$$= \frac{\text{rise}}{\text{run}} \tag{97}$$

$$= \frac{(0 \text{ mph}) - (+300 \text{ mph})}{15 \text{ s} - 12 \text{ s}} \tag{98}$$

$$= -100 \frac{\text{mph}}{\text{s}}. \tag{99}$$

What Does a Negative Value of a_x Mean?

In the last Worked Example, a_x was negative, and the dragster was slowing down. So it's natural to wonder: Does slowing down go hand in hand with negative a_x values?

The answer is *no!* A negative a_x value can correspond to slowing down *or* speeding up. In fact, *you cannot even tell* from a_x alone whether an object is speeding up or slowing down!

The fact is, you have to look at the acceleration vector *and* the velocity vector to know how an object is moving. Remember Chapter 6. If the acceleration vector points *along* the velocity vector, the object is speeding up. If the acceleration vector points *against* the velocity vector, the object is slowing down. You need both **v** and **a** to decide.

Now would be a good time to try Focused Problem 7.

By the way, the ninth row of Table 3 corresponds to a situation in which a_x is negative for an object that is speeding up.

When a_x Remains Constant over Time

In Worked Example 4, we sketched the $a_x - t$ graph for the dragster. The graph is reproduced here for convenience (Figure 133).

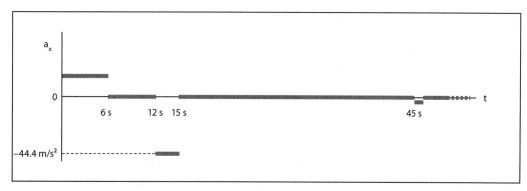

FIGURE 133. The $a_x - t$ graph for the dragster in Worked Example 4.

Focused Problems 8–13 deal with the rest of the material in this chapter.

As you can see, the x-component of the acceleration vector changes over time. For a while a_x is positive, then a_x drops to zero, then for a while a_x is negative, then a_x goes back to zero, then a_x is briefly negative, and then a_x goes back to zero.

Generally, the acceleration of a moving object *will* vary with time in some way. But there are idealized situations in which a_x remains constant over time. For example, in the case of the dragster, if we focus on the interval from $t = 0$ to $t = 6$ s, we have a situation in which a_x remains constant.

Some specialized formulas are available for use in the case of constant a_x. Instead of giving a stock derivation of these formulas, I'll try to give you some deeper insights by working up to the formulas graphically using what we already know about rates of change.

Later on, this Worked Example will prove to be crucial for understanding Newton's System of the World and the entire deterministic worldview.

»WORKED EXAMPLE 9

An object moving along the x-axis has its $a_x - t$ graph shown in Figure 134.

At $t = 0$, it is known that the object is located at horizontal position $x = +6$ m and has horizontal velocity component $v_{0,x} = -12$ m/s. Use the given information at $t = 0$, together with the given $a_x - t$ graph, to sketch $v_x - t$ and $x - t$ graphs for the entire motion. Align your graphs top to bottom, with the $x - t$ graph on top, the $v_x - t$ graph in the middle, and the $a_x - t$ graph on the bottom.

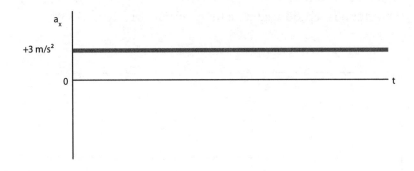

FIGURE 134.
Worked Example 9. The $a_x - t$ graph of an object moving along the x-axis.

»SOLUTION 9

In Figure 135 I have set up some graph paper so that we can draw our three graphs. The $a_x - t$ graph was given in the problem, so I just placed it at the bottom. I have also represented the given

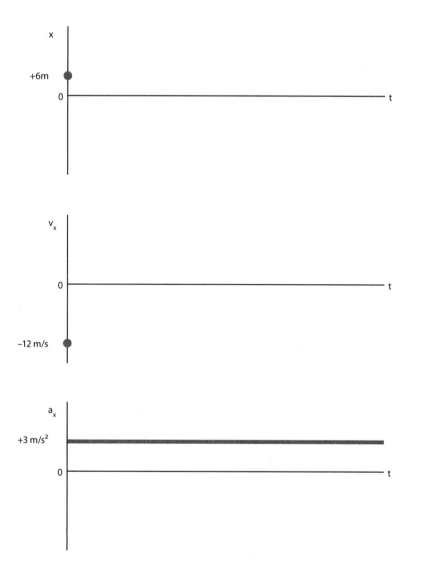

FIGURE 135.
Solving Worked
Example 9. Getting
ready to draw the
$v_x - t$ and $x - t$
graphs for the entire
motion depicted in
Figure 134.

information at $t = 0$ by drawing dots on the $v_x - t$ graph and the
$x - t$ graph.

The strategy here is to work our way up the page one graph
at a time, moving from bottom to top. So let's focus on drawing
the $v_x - t$ graph. We have to get from the $a_x - t$ graph to the $v_x - t$
graph. Notice that this is the reverse of what we usually do! Usually
we start from the $v_x - t$ graph and sketch the $a_x - t$ graph by
eyeballing the slope of the $v_x - t$ graph at each point. But in this
case, we have to start with $a_x - t$ and go backward.

Here's how we go backward, then. If we look at the $a_x - t$ graph,
we can see that it shows a *constant positive value* over time. Therefore,

the $a_x - t$ graph is telling us to draw a $v_x - t$ graph with a *constant positive slope* over time. What kind of curve has a constant positive slope? A straight line directed upward. So our $v_x - t$ graph must be a straight line directed upward.

Unfortunately, there are many ways to draw a straight line directed upward. Some of the possibilities are shown in Figure 136. All of these are consistent with the slope information in the given a_x graph.

This is where the initial conditions come in. At time $t = 0$, v_x was given as -12 m/s. So the intercept of our $v_x - t$ graph must be at -12 m/s. This leads us to pick out Figure 136(d) as the kind of

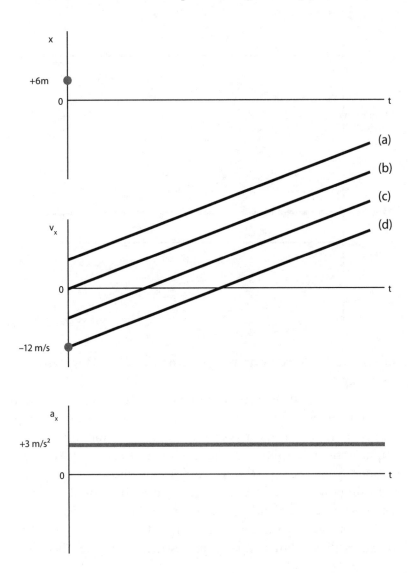

FIGURE 136.
Solving Worked Example 9. Some possible $v_x - t$ graphs for the motion depicted in Figure 134.

$v_x - t$ graph we want. I have sketched this into our graph paper in Figure 137.

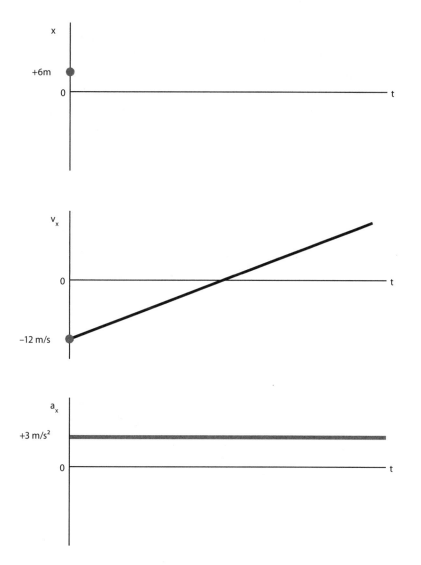

FIGURE 137.
Solving Worked
Example 9. The $v_x - t$
curve that matches
the given initial
condition at $t = 0$
and has a slope at
every moment equal
to the value of a_x at
that moment.

Now that we've gone from a_x to v_x, we can keep moving up the ladder and use the same kind of trick to get from v_x to x.

If we look at the $v_x - t$ graph, we can see that it charts a value that *starts out negative, becomes zero, and then becomes more and more positive.* What we need to do, then, is draw an $x - t$ graph for which the *slope* starts out negative, becomes zero, and then becomes more and more negative.

The $v_x - t$ graph was uniquely determined by two things: (i) the slope information in the $a_x - t$ graph, and (ii) the initial value of v_x.

In simpler terms, we need the $x - t$ curve to start out steeply downhill, gradually flatten out horizontally, and then gradually become steeper and steeper uphill. In hiking terms, what we've just described is a valley.

So the $x - t$ curve looks like a valley. But again, there are many ways to sketch a valley, even if we agree on a basic shape. Some of the possibilities are shown in Figure 138. All of these are consistent with the slope information in the given $v_x - t$ graph.

This is where the initial conditions come in again. At time $t = 0$, x was given as +6 m. So the intercept of our $x - t$ graph must be

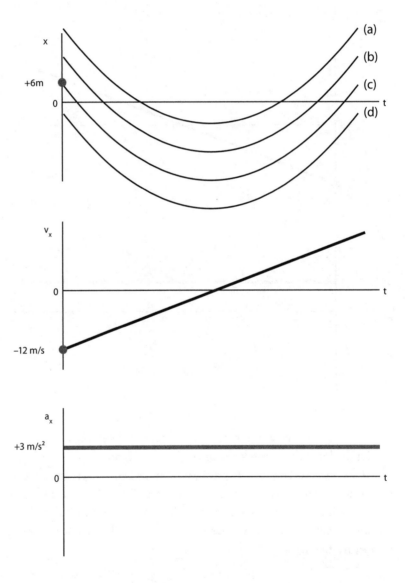

FIGURE 138.
Solving Worked Example 9. Some possible $x - t$ graphs.

at $+6$ m. This leads us to pick out Figure 138(c) as the kind of $x - t$ graph we want. I have sketched this onto our graph paper as Figure 139. Now we are done. ◇

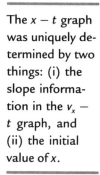

The $x - t$ graph was uniquely determined by two things: (i) the slope information in the $v_x - t$ graph, and (ii) the initial value of x.

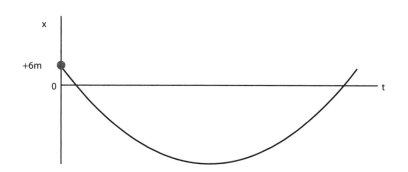

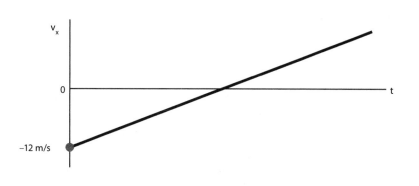

FIGURE 139. Solution to Worked Example 9. The $x - t$ curve that matches the given initial condition at $t = 0$ and has a slope at every moment equal to the value of v_x at that moment, along with the $v_x - t$ curve and the $a_x - t$ curve we determined earlier.

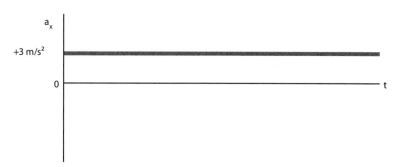

Acting Out Motion with Constant Acceleration

What would the motion in Figure 139 look like if you acted it out?

Actually, it's tricky to think about acting out this motion, because in this motion you have to be moving already at time $t = 0$. So to act it out properly, you really have to think about it almost as if it's a scene in a movie—the kind of scene where the action is already in progress when the director cuts to the scene. In

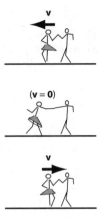

FIGURE 140.
A dancer's changing
velocity vector.

other words, you have to get yourself moving *before* $t = 0$, then yell "*Action!*" when you're ready to "start filming" at $t = 0$.

When we join the "scene" diagrammed in Figure 139, the actor is located 6 meters to the right of the origin, already in motion at 12 meters per second to the left, and already in the process of slowing down at a rate of 3 meters per second per second. As the camera rolls, the actor continues to slow until she momentarily comes to a stop and reverses her direction, now moving to the right and gradually speeding up as she goes. This could be a Lindy Hop scene in which the dancer is being flung outward and drawn back in one fluid motion.

The dancer's velocity vector starts off pointing to the left, gradually shrinks away to nothing, then emerges again pointing to the right, and gradually grows in length. The turn-around portion of the motion is illustrated in Figure 140.

Although the dancer's velocity vector changes with time, her acceleration vector always points to the right, with a constant magnitude of 3 m/s². This is shown in Figure 141.

Notice that one unchanging acceleration vector describes *both* the dancer's slowing-down phase *and* her speeding-up phase! This works out because in the first half of the motion the acceleration vector points *against* the velocity vector, slowing her down, and in the second half of the motion the acceleration vector points *with* the velocity vector, speeding her up. It's **v** that changes here, not **a**.

Equations of Motion when a_x Is Constant

You probably noticed that the $x - t$ curve in Worked Example 9 looked like a parabola. Indeed, as long as a_x is constant and

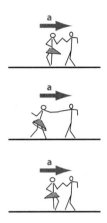

FIGURE 141.
The dancer's constant acceleration vector.

nonzero, the position x will vary quadratically with time and will therefore graph as a parabola. This is the meaning of the $\frac{1}{2}at^2$ formulas, which I will present next.

It takes a bit of work just to write out the formulas, because we have to get our symbols straight.

In Worked Example 9 we saw that if you know the constant value of an object's acceleration a_x and if you know the object's initial position and initial velocity, you can determine the object's position and velocity at later times. So in general, think of the following quantities as the "givens":

- x_0 is the x-coordinate of the position vector at time $t = 0$.
- $v_{0,x}$ is the x-component of the velocity vector at time $t = 0$.
- a_x is the x-component of the acceleration vector, which we are here assuming remains constant over time.

Now think of the following quantities as what we want to know, in terms of the givens:

- $x(t)$ is the x-coordinate of the position vector at any later time t.
- $v_x(t)$ is the x-component of the velocity vector at any later time t.

Without further ado, the formulas for the sought-after quantities are as follows:

$$x(t) = x_0 + v_{0,x}t + \frac{1}{2}a_x t^2 \qquad (100)$$

$$v_x(t) = v_{0,x} + a_x t. \qquad (101)$$

For now, I'll just quote the formulas and focus on asking some helpful questions about them. In Focused Problem 13 I'll lead you through your own derivation of the formulas.

If you know
calculus, take a
moment to verify
that $v_x(t) = x'(t)$.

Both of these formulas have easy graphical interpretations. Let's start with velocity, for which the formula is $v_x(t) = v_{0,x} + a_x t$. If we were to graph $v_x(t)$ as a function of time, in general the relation $v_x(t) = v_{0,x} + a_x t$ would yield a *straight line*, with slope equal to a_x and intercept equal to $v_{0,x}$. In general, the line would have some positive or negative slope, although in the one special case $a_x = 0$, the line for $v_x(t)$ would be horizontal at the constant value $v_{0,x}$.

Now for position, for which the formula is $x(t) = x_0 + v_{0,x}t + \frac{1}{2}a_x t^2$. If we were to graph $x(t)$ as a function of time, in general the relation $x(t) = x_0 + v_{0,x}t + \frac{1}{2}a_x t^2$ would yield a *parabolic curve*. With a_x positive, the parabola would open upward. With a_x negative, the parabola would open downward.

In the special case $a_x = 0$, the graph wouldn't be a parabola at all: instead the curve would degenerate into a straight line. In this case, the straight line $x(t) = x_0 + v_{0,x}t$ would describe an object tracking along the x-axis at a constant rate (such as the clown in Worked Example 3 of Chapter 7).

In the super-special case $a_x = 0$ *and* $v_{0,x} = 0$, the straight line would further degenerate into a *horizontal* line, $x(t) = x_0$. This is the $x - t$ graph for an object that is sitting still.

» WORKED EXAMPLE 10

A ball is tossed up into the air, as by a tennis player preparing for a serve. By the time we join the scene at $t = 0$, the ball is already at a height 2 m above the ground, moving upward at a speed of 20 m/s and slowing at a constant rate of 10 m/s². Assume that the ball's acceleration vector remains constant throughout the motion, pointing downward with magnitude 10 m/s². See Figure 142.

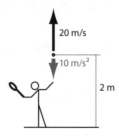

FIGURE 142.
Worked Example 10.
A ball being tossed
up in the air; the
situation at time
$t = 0$.

(i) Taking the y-axis to be pointing upward, with the origin at ground level, sketch $a_y - t$, $v_y - t$, and $y - t$ graphs for the ball's motion. Extend your graphs to the time when the ball returns to height 2 m. (ii) Write an expression for the y-velocity of the ball $v_y(t)$ at any time t during the motion. Use all of the relevant given information. (iii) Write an expression for the y-position of the ball $y(t)$ at any time t during the motion. Use all of the relevant given information. (iv) How high is the ball at time $t = 3$ s? (v) How high will the ball get? (vi) At what two times is the ball moving at a speed of 6 m/s? (vii) Sketch the velocity and acceleration vectors of the ball at these two times. (viii) Johnny thinks that the acceleration of the ball is zero when the ball is at its peak. Give five reasons why Johnny can't be right.

» SOLUTION 10

i. I have laid down the requested coordinate system in Figure 143.

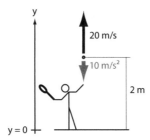

FIGURE 143. Solving Worked Example 10. A coordinate system for the problem.

Relative to this coordinate system, my sketches of a_y, v_y, and y versus time are shown in Figure 144.

Notice that my a_y curve maintains a constant value of -10 m/s² throughout the motion. The v_y curve starts at the appropriate value of $+20$ m/s and has a constant negative slope, as dictated by the $a_y - t$ graph. The y curve starts at the appropriate value of $+2$ m, and the slope of the $y - t$ curve is initially steep uphill, as dictated by the positive value of the initial velocity v_y. The slope gradually flattens out, as dictated by the decreasing speed shown in the v_y curve. At the moment when the $v_y - t$ curve crosses the t-axis, the ball is momentarily at rest and the $y - t$ curve is momentarily flat. (Notice from Figure 144 that nothing special is going on in the $a_y - t$ graph at this moment.) As the v_y curve drops below the t-axis, the $y - t$ curve becomes steeper and steeper downhill. The scene is over when the position returns to $y = +2$ m.

ii. To find the y-velocity of the ball as a function of time, start with the stock formula. This formula is really just a slogan, so I

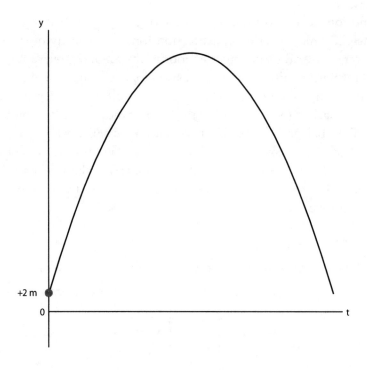

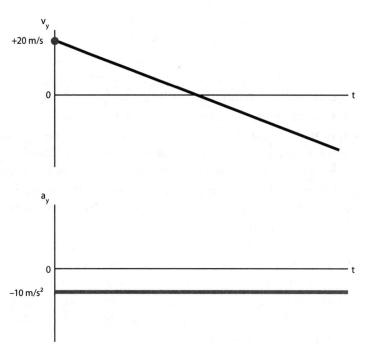

FIGURE 144. Solving Worked Example 10. Graphs of a_y, v_y, and y versus time for a ball thrown upward.

put it in quotation marks. Then go through and specialize to the current problem using the given initial conditions.

$$\text{``}v_y(t) = v_{0,y} + a_y t\text{''} \tag{102}$$

$$v_y(t) = (+20 \text{ m/s}) + (-10 \text{ m/s}^2)t \tag{103}$$

$$v_y(t) = 20 \text{ m/s} - (10 \text{ m/s}^2)t. \tag{104}$$

Equation 104 gives the y-velocity of the ball at any given time during the scene.

If you graph $v = 20 - 10t$ on a graphing calculator, you will see the straight-line graph that we sketched earlier for v_y as a function of time.

---

    iii. To find the y-position of the ball as a function of time, we again start with the appropriate stock formula and specialize it with the given initial conditions:

$$\text{``}y(t) = y_0 + v_{0,y}t + \frac{1}{2}a_y t^2\text{''} \tag{105}$$

$$y(t) = (+2 \text{ m}) + (+20 \text{ m/s})t + \frac{1}{2}(-10 \text{ m/s}^2)t^2 \tag{106}$$

$$y(t) = 2 \text{ m} + (20 \text{ m/s})t - (5 \text{ m/s}^2)t^2. \tag{107}$$

Equation 107 gives the y-position of the ball at any time during the scene.

If you graph $y = 2 + 20t - 5t^2$ on a graphing calculator, you will see the parabolic curve that we sketched above for y as a function of time.

    With $y(t)$ and $v_y(t)$ known explicitly as functions of time, there's nothing about the motion we can't figure out—as we'll see next.

    iv. We can find the vertical position of the ball at time $t = 3$ s by evaluating our $y(t)$ formula at that time:

$$y(t) = 2 \text{ m} + (20 \text{ m/s})t - (5 \text{ m/s}^2)t^2 \tag{108}$$

$$y(3 \text{ s}) = 2 \text{ m} + (20 \text{ m/s})(3 \text{ s}) - (5 \text{ m/s}^2)(3 \text{ s})^2 \tag{109}$$

$$y(3 \text{ s}) = 2 \text{ m} + 60 \text{ m} - (5 \text{ m/s}^2)(9 \text{ s}^2) \tag{110}$$

$$y(3 \text{ s}) = 2 \text{ m} + 60 \text{ m} - 45 \text{ m} \tag{111}$$

$$y(3 \text{ s}) = +17 \text{ m.} \tag{112}$$

At time $t = 3$ s, the vertical position of the ball is $y = +17$ m. By itself, this doesn't tell us how high the ball is. But it does tell us that the ball is a distance of 17 m to the arrowhead side of the origin, and, given the way we laid down our coordinate system, this does in fact mean that the ball is 17 m above the ground at this point.

v. Here is one way to find the maximum height of the ball. The ball's maximum height occurs at the moment in time when its y-velocity is zero. (Can you see this from the $y - t$ and $v_y - t$ graphs?) So we can use our formula for $v_y(t)$ to find the time t_{peak} when v_y reaches 0. To do this, set $v_y = 0$ on the left-hand side, and solve for t:

$$(113) \qquad v_y(t) = 20 \text{ m/s} - (10 \text{ m/s}^2)t$$

$$(114) \qquad 0 = 20 \text{ m/s} - (10 \text{ m/s}^2)t_{peak}$$

$$(115) \qquad (10 \text{ m/s}^2)t_{peak} = 20 \text{ m/s}$$

$$(116) \qquad t_{peak} = \frac{(20 \text{ m/s})}{(10 \text{ m/s}^2)} = 2 \text{ s}.$$

If you didn't need to use any formulas to find t_{peak}, good for you. The velocity starts off at $+20$, and it's decreasing at a steady rate of 10 every second, so it's naturally going to take 2 seconds to fall to zero.

Knowing t_{peak}, we can now simply evaluate the y-position at this time:

$$(117) \qquad y(t) = 2 \text{ m} + (20 \text{ m/s})t - (5 \text{ m/s}^2)t^2$$

$$(118) \qquad y(2 \text{ s}) = 2 \text{ m} + (20 \text{ m/s})(2 \text{ s}) - (5 \text{ m/s}^2)(2 \text{ s})^2$$

$$(119) \qquad y_{peak} = 22 \text{ m}.$$

The peak height is therefore 22 m (given the way we laid down our coordinate system). We see now that in part (iv) when we found $y = 17$ m at $t = 3$ s, we were looking at the ball on its way down.

This would be a pretty intimidating tennis serve—especially coming from a player who looks to be about four feet tall.

If you know calculus, another way to find the peak height is to take the derivative of $y(t)$ and set it equal to zero. You should find that the derivative of $y(t)$ is the same expression we are using for $v_y(t)$. This means that the two strategies are really the same.

vi. The speed of the ball will be 6 m/s when $v_y = +6$ m/s or $v_y = -6$ m/s. We can use our formula for $v_y(t)$ to find the times when these things happen:

$$v_y(t) = 20 \text{ m/s} - (10 \text{ m/s}^2)t \tag{120}$$

$$\pm 6 \text{ m/s} = 20 \text{ m/s} - (10 \text{ m/s}^2)t \tag{121}$$

$$-14 \text{ m/s} = -(10 \text{ m/s}^2)t \tag{122}$$

or

$$-26 \text{ m/s} = -(10 \text{ m/s}^2)t \tag{123}$$

$$t = 1.4 \text{ s} \tag{124}$$

or

$$t = 2.6 \text{ s.} \tag{125}$$

These two times are indicated on the $y - t$, $v_y - t$, and $a_y - t$ graphs in Figure 145, along with the rest of our quantitative results. At $t = 1.4$ s, the ball is on the way up. At $t = 2.6$ s, the ball is on the way down. At time $t = 2$ s, the ball is momentarily at rest at its peak height of 22 m above the ground. At $t = 3$ s, the ball is at height 17 m, on the way down, and moving faster than 6 m/s.

vii. My sketches of the velocity and acceleration vectors of the ball are shown in Figure 146. Notice that the acceleration vector is the same at both instants of time shown. (The ball is like the dancer in Figures 140 and 141.)

viii. Here are five reasons—plus a silly one—why Johnny is wrong in thinking that the acceleration of the ball is zero when the ball is at its peak:

1. First the silly reason: This problem is in a section on *constant* acceleration, so it would be very strange if the acceleration suddenly vanished and reappeared again during the scene.

2. Now for a somewhat less silly reason: The problem *says* that the acceleration stays constant, so the acceleration can't suddenly vanish and reappear during the scene.

3. The graph of a_y versus time *shows* that the acceleration is not zero at the peak!

4. It is quite possible for the ball to be accelerating without moving, if only for a moment. The acceleration vector tells what is *about to happen* to the velocity vector, and at the peak the velocity vector is *about* to point downward. Hence, the

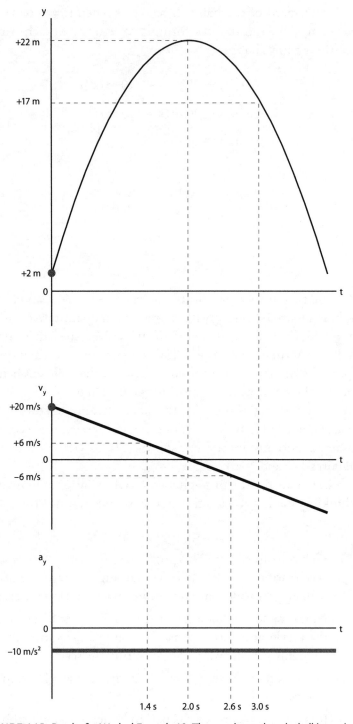

FIGURE 145. Results for Worked Example 10. The two times when the ball is moving at 6 m/s.

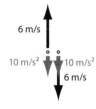

FIGURE 146.
Results for Worked
Example 10. The
velocity and acceler-
ation vectors for the
ball at $t = 1.4$ s and
$t = 2.6$ s. For clarity,
the ball is shown dis-
placed slightly to the
right at the later time.

acceleration vector points downward at the peak. (Compare this with Figure 96 of Chapter 6.)

5. The slope of the $v_y - t$ graph is clearly negative (downhill), not zero, at the moment when it passes through the t-axis.

6. The value of v_y just before the peak is positive, and the value of v_y just after the peak is negative, so obviously v_y was in the process of changing during this entire time.

OK, that's six, and I'm out of reasons. ◇

"Collision Problems"

What good would studying physics be if you couldn't use it to solve those pesky word problems?

»WORKED EXAMPLE 11

Alice leaves San Francisco headed east on Interstate 80 at 70 mph. At the same time, 3,000 miles away, Bob leaves New York City headed west on Interstate 80 at 65 mph. When and where will they pass each other?

(Here we're making a host of unrealistic assumptions—that I-80 is a straight line, that the two drivers never stop or even change speed, and so on. I just want to illustrate a general approach using this simplified situation.)

»SOLUTION 11

If you'd never studied physics, you might approach this problem in the following way. You might say, well, the distance traveled by

Alice and the distance traveled by Bob have to add up to 3,000 miles, and the distances they travel are going be in the ratio of 70 to 65 because that's the ratio of their speeds. This leads to the two equations $d_A + d_B = 3,000$ and $d_A/d_B = 70/65$, which can be solved to find d_A and d_B.

This is fine, and it works for the problem at hand, but you will *not* be able to apply simple proportional reasoning like this to the situations you will find in the Focused Problems. So let's take this opportunity to learn a strategy that generalizes.

As always, we are going to start by drawing a cartoon and laying down a coordinate system. My choices are shown in Figure 147.

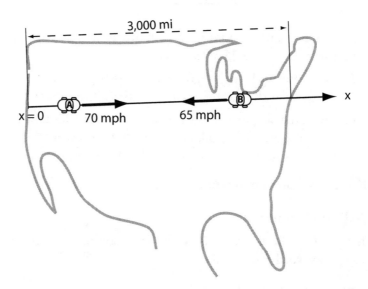

FIGURE 147.
Worked Example 11. A cartoon and coordinate system for solving the problem.

Now that we have a coordinate system, we can sketch graphs of position versus time for Alice and Bob. Let's start with Alice.

Alice starts from the origin ($x = 0$) and moves with the arrowhead at a constant rate. So her $x - t$ graph is a straight line with a positive slope. This is shown in Figure 148.

Bob, meanwhile, starts from position $x = +3,000$ mi and moves against the arrowhead at a constant rate. So his $x - t$ graph is a straight line with positive intercept and negative slope. This is shown in Figure 149. In this figure I have overlaid the two graphs, because now it is clear what's going to happen: When the two curves cross, that's the moment when Alice and Bob will meet.

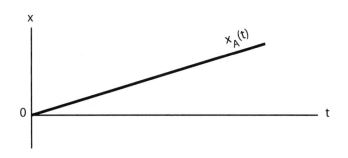

FIGURE 148.
Solving Worked
Example 11. Alice's
$x - t$ graph.

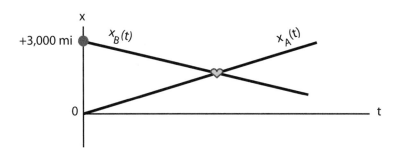

FIGURE 149.
Solving Worked
Example 11. Bob's
$x - t$ graph overlaid
on Alice's $x - t$
graph.

To analyze the problem quantitatively, we'll proceed just as we did in Worked Example 10, by specializing the stock formulas to the problem at hand. For the position of Alice, we have

$$\text{``}x(t) = x_0 + v_{0,x}t + \frac{1}{2}a_x t^2\text{''} \tag{126}$$

$$x_A(t) = (0) + (+70\ \text{mph})t + \frac{1}{2}(0)t^2 \tag{127}$$

$$x_A(t) = (70\ \text{mph})t. \tag{128}$$

This gives the x-position of Alice at any time during the scene.

Now for the position of Bob:

$$\text{``}x(t) = x_0 + v_{0,x}t + \frac{1}{2}a_x t^2\text{''} \tag{129}$$

$$x_B(t) = (3{,}000\ \text{mi}) + (-65\ \text{mph})t + \frac{1}{2}(0)t^2 \tag{130}$$

$$x_B(t) = 3{,}000\ \text{mi} - (65\ \text{mph})t. \tag{131}$$

This gives the x-position of Bob at any time during the scene.

Alice and Bob will meet when they occupy the same point on the number line—that is, when their positions are the same. So to find the time when they will meet, we demand that their positions be equal:

$$x_A(t) = x_B(t) \tag{132}$$

$$(70 \text{ mph})t = 3{,}000 \text{ mi} - (65 \text{ mph})t. \tag{133}$$

Now it is easy to solve for their meeting time, t_{meet}:

$$(70 \text{ mph})t_{meet} = 3{,}000 \text{ mi} - (65 \text{ mph})t_{meet} \tag{134}$$

$$(70 \text{ mph})t_{meet} + (65 \text{ mph})t_{meet} = 3{,}000 \text{ mi} \tag{135}$$

$$(135 \text{ mph})t_{meet} = 3{,}000 \text{ mi} \tag{136}$$

$$t_{meet} = \frac{3{,}000 \text{ mi}}{135 \frac{\text{mi}}{\text{hr}}} \tag{137}$$

$$t_{meet} = 22.2 \text{ hr.} \tag{138}$$

At this time, where will Alice and Bob be located along the coordinate axis? We can plug t_{meet} into x_A or x_B to find out:

$$x_{meet} = x_A(t_{meet}) \tag{139}$$

$$= 70 \text{ mph} \times 22.2 \text{ hr} \tag{140}$$

$$x_{meet} = 1{,}554 \text{ mi.} \tag{141}$$

We could also have plugged t_{meet} into x_B and found the same answer, because we determined t_{meet} in the first place by requiring that x_A and x_B have the same value at that time.

So the meeting will occur at time $t_{meet} = 22.2$ hr, at position $x = 1{,}554$ mi. This information is shown on the $x - t$ graph in Figure 150.

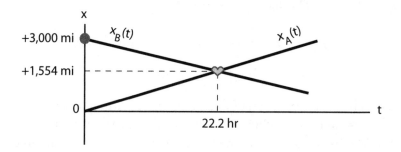

FIGURE 150.
Solution to Worked Example 11, showing where Alice and Bob will meet.

With reference to the coordinate system, the meeting point is 1,554 miles east of San Francisco, 1,446 miles west of New York City, and 54 miles east of the midpoint. The meeting point will be slightly east of the midpoint because Alice is driving slightly faster than Bob. ◇

This last Worked Example is what I call a "collision problem." In a collision problem, you want to know when two objects will collide, when two objects will meet, or when one object will catch up with another object. I always use the same strategy for collision problems. Here's the method:

1. Based on the stock formula $x(t) = x_0 + v_{0,x}t + \frac{1}{2}a_x t^2$, write down the position of object A as a function of time, $x_A(t)$. Use given information to specify the values of x_0, $v_{0,x}$, and a_x as far as possible. But leave the t alone.

2. Similarly, write down the position of object B as a function of time, $x_B(t)$.

3. Set $x_A(t) = x_B(t)$, and solve for t. This is the collision time t_{coll}. (Assuming that a collision does indeed occur!)

4. Now you can insert t_{coll} into $x_A(t)$ or $x_B(t)$ to find the common position of object A and object B when the collision occurs.

5. You can also insert t_{coll} into $v_{A,x}(t)$ and $v_{B,x}(t)$ if you want to know the (generally different) x-velocities of objects A and B at the collision time.

Quick strategy for collision problems:

- Write out $x_A(t)$,
- Write out $x_B(t)$,
- Set $x_A(t) = x_B(t)$ and solve for t.

Reader's Notes

Below is space for you to summarize key points from this chapter.

FOCUSED PROBLEMS :: CHAPTER 9

1. Last night I watched my cat Peanut as she walked on a windowsill in my living room. The graph in Figure 151 illustrates the movement that I observed. As suggested by the numbers shown in the cartoon, the origin is at the center of the windowsill, and the x-arrowhead points to the right.

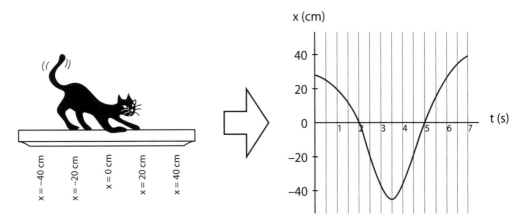

FIGURE 151. Focused Problem 1. My cat Peanut walking on a windowsill and an x − t graph of her movement.

a. Judging from the x − t graph, when, if ever, is Peanut located to the left of the origin?

b. When, if ever, is Peanut moving leftward?

c. When, if ever, is Peanut moving rightward?

d. When, if ever, is Peanut's x-velocity equal to zero?

e. Sketch a careful, non-numerical graph of Peanut's x-velocity as a function of time.

f. When, if ever, is Peanut speeding up?

g. When, if ever, is Peanut slowing down?

h. Write a descriptive account of Peanut's motion from start to finish in your own words. Act out the motion yourself, and see if you and your study partners agree with one another.

2. At time $t = 0$, a dragster starts from rest and gradually speeds up at a constant rate, reaching a top speed of 300 mph at time $t = 6$ s. Compute the x-acceleration during this initial speed-up phase. Label Figure 133 with this information.

3. Figure 152 shows the $x - t$ graph for an electron moving along the x-axis. Carefully estimate the numerical value of v_x at time $t = 3.5$ s. Also describe this entire motion sequence in words, as if you were watching it in a movie. Where does the electron start? How fast is it going at first? When the scene begins at $t = 0$, is the electron in the process of speeding up? Slowing down? Neither? How fast does the electron end up moving? In which direction? Does it ever turn around or switch directions?

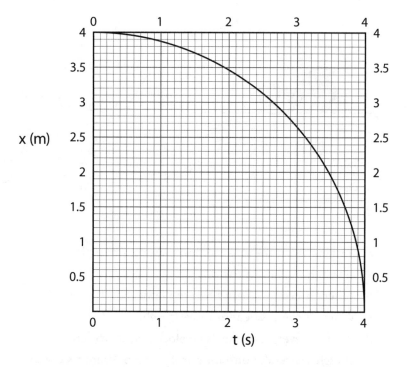

FIGURE 152.
Focused Problem 3.
The $x - t$ graph for an electron moving along the x-axis.

4. Figure 153 shows the $x - t$ graph for a car driving along the x-axis. Sketch the $v_x - t$ and $a_x - t$ graphs for this motion. Describe the motion in words, and act it out. Is the horizontal acceleration of the car constant in time?

5. Figure 154 shows the $v_x - t$ graph for a taxicab driving from one stoplight to another, moving along the x-axis. In the spaces provided, sketch the $a_x - t$ graph as well as one possible $x - t$ graph. Describe the motion in words, and act it out. How would a cab driver operate the pedals to achieve a velocity profile like the one shown in Figure 154?

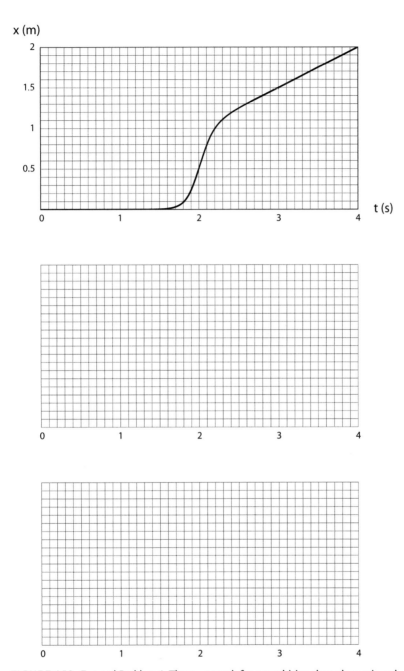

FIGURE 153. Focused Problem 4. The $x - t$ graph for a car driving along the x-axis and coordinates for you to use in drawing the $v_x - t$ and $a_x - t$ graphs.

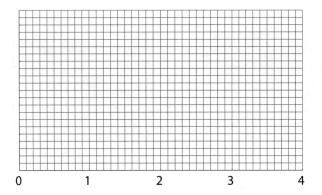

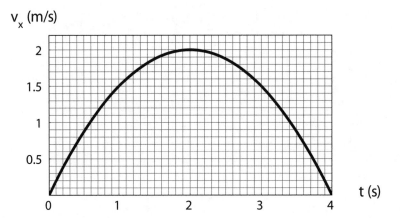

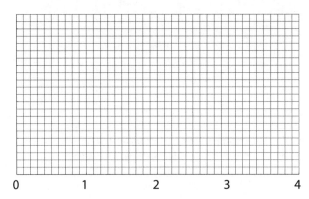

FIGURE 154. Focused Problem 5. The $v_x - t$ graph for a taxicab driving from one stoplight to another and coordinates for you to use in drawing the $a_x - t$ graph and one possible $x - t$ graph.

6. A proton is fired into a plasma; forces from the plasma slow the proton until the proton comes to a stop. Gravity is negligible in this problem; all of the motion is horizontal. The initial horizontal velocity of the proton is 10^5 m/s. The proton enters the plasma at $t = 0$, and its velocity profile thereafter is given by the formula

$$v_x(t) = 10^5 \text{ m/s}\sqrt{1 - (t/T)^2}, \tag{142}$$

where T is a constant with the value $T = 10^{-6}$ s. (This is an idealized example.)

   a. According to Equation 142, what is the x-velocity of the proton at time $t = 0$? What is the x-velocity of the proton at time $t = 5 \times 10^{-7}$ s? What is the x-velocity of the proton at time $t = 10^{-6}$ s?

   b. Sketch a $v_x - t$ graph for the proton over the period from $t = 0$ to $t = 10^{-6}$ s. (Hint: We can rearrange Equation 142 to read $(v_x/10^5 \frac{\text{m}}{\text{s}})^2 + (t/T)^2 = 1$, so the $v_x - t$ curve is a quarter of an ellipse.)

   c. Sketch an $a_x - t$ graph for the proton.

   d. Describe the proton's motion in words.

7. With reference to the proton in Focused Problem 6, use Equation 84 to calculate how far the proton travels during its motion. (Hint: The area A of an ellipse with semimajor axis a and semiminor axis b is given by $A = \pi a b$.)

8. At $t = 0$, Agent Smith throws an unconscious Trinity straight down from the top of a tall building of height 300 m with an initial downward velocity of 10 m/s. Assume that Trinity's acceleration vector always points straight down, with a magnitude of 9.8 m/s^2 that remains constant over time.

   Neo, who until this time has been hovering directly above Trinity at a height of 600 m above the ground, immediately streaks downward, speeding up from rest at a constant rate of 29.4 m/s^2. *Will Neo catch Trinity before it's too late?*

   a. Draw a cartoon of this situation, including a coordinate system.

b. Sketch a graph of Trinity's and Neo's positions as a function of time on the same set of axes. Of course, without solving the problem in detail, you can't really draw a specific graph. So sketch graphs that show two scenarios: (i) Neo does in fact catch up to Trinity before it's too late and (ii) Neo does not catch up to Trinity in time.

c. Find the position at which Neo catches up to Trinity. How do you interpret this result?

9. A cheetah is hunting a wildebeest on the African plains. At time $t = 0$, the cheetah is $D = 100$ m behind the wildebeest, closing in at speed $v_0 = 30$ m/s. The cheetah maintains this constant speed throughout the chase.

At this same instant ($t = 0$), the wildebeest first notices the cheetah and begins speeding up from rest at a constant rate of $a_0 = 5$ m/s^2 in an effort to escape the cheetah. See Figure 155.

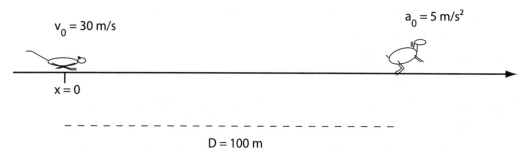

$v_0 = 30$ m/s

$a_0 = 5$ m/s^2

x = 0

D = 100 m

FIGURE 155.
Focused Problem 9.
A cheetah and
wildebeest at $t = 0$.

Throughout this problem, you can give your answers numerically or in terms of the given symbols.

a. Those of you rooting for the wildebeest will be pleased to know that the numbers given earlier are such that the cheetah does *not* catch the wildebeest. Knowing this, sketch accurate but non-numerical graphs of the positions of both the cheetah and the wildebeest as functions of time on a single $x - t$ graph. Make your sketches with reference to the origin and positive direction that are indicated in the Figure 155.

b. Find the time at which the cheetah and the wildebeest have the same x-velocity.

c. At the moment when the cheetah and the wildebeest have the same x-velocity, how far ahead of the cheetah is the wildebeest?

d. On your graph from part (a), indicate the moment in time when the cheetah and the wildebeest have the same x-velocity. How did you decide where to put your mark?

e. The wildebeest gets away because its head start is so large ($D = 100$ m). If its head start had been smaller, the cheetah would have caught her prey. What is the critical value for the head start that divides escape from capture?

10. Use the constant acceleration formulas for $x(t)$ and $v_x(t)$ to derive the formula

$$v_{f,x}^2 - v_{i,x}^2 = 2a_x\Delta x. \qquad (143)$$

This formula comes in very handy when you're solving a constant acceleration problem in which you aren't given any time data.

11. A plane must attain a certain minimum speed in order to take flight. Let's say that this minimum speed is 120 m/s. We'll suppose that the plane starts from rest at the beginning of a 1,000-m runway and speeds up at a constant rate.

a. Draw $x - t$, $v_x - t$, and $a_x - t$ graphs for the plane.

b. What minimum acceleration must the plane have in order to achieve takeoff speed before it reaches the end of the runway?

c. After solving the problem, label the graphs quantitatively. Also, make a table showing the plane's x-velocity at each of the times indicated in Figure 156.

12. Alice and Bob are staying at a hotel in Des Moines before turning around and driving back to their home cities. (Recall Worked Example 11 of this chapter.) To freshen up after a long day of driving, they decide to go swimming in the hotel's dinky 10-m pool.

t	$v_x(t)$
0	
1 s	
2 s	
3 s	
4 s	
...	
15 s	
16 s	
17 s	

FIGURE 156.
Focused Problem 11.
Times in a plane's
takeoff.

Bob challenges Alice to a race from one end of the pool to the other. In fact, Bob is so confident of victory that he lets Alice use the starting block, whereas he himself will start from the wall. Alice is no fool and takes him up on his offer.

At $t = 0$, Alice hits the water 2 m out with a horizontal velocity of 1 m/s. She maintains her speed throughout the race. Meanwhile, at $t = 0$ Bob starts off from rest at the wall, speeding up at a constant rate of 0.5 m/s^2. See Figure 157.

At $t = 0$:

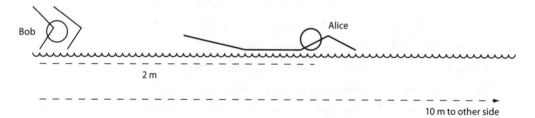

FIGURE 157.
Focused Problem 12.
Bob and Alice racing
across a swimming
pool.

a. Sketch rough, non-numerical $x - t$ graphs for Alice and Bob on the same set of axes.

b. Who wins the race?

Bob is not in the best shape of his life. So the assumption that he can keep increasing his speed at a steady rate throughout the race is, shall we say, unrealistic. So instead let us suppose that he speeds up from rest at 0.5 m/s^2 until

his speed reaches 1.5 m/s and that he maintains this speed
for the remainder of the race.

c. Sketch a new $x - t$ curve for Bob, also showing Alice's $x - t$
   curve on the same set of axes.

d. At what time does Bob reach his maximum speed? Where
   is Bob by this point? Where is Alice?

e. Who wins the race under this set of assumptions?

13. The goal of this problem is to derive the formula $x(t) = x_0 + v_{0,x}t + \frac{1}{2}a_x t^2$ for motion with constant a_x. To get ready for this,
we'll express the formula in the following way:

$$x_f - x_i = v_{i,x}\left(t_f - t_i\right) + \frac{1}{2}a_x\left(t_f - t_i\right)^2 . \tag{144}$$

We'll derive the formula using our rule for net changes from
Chapter 2:

$$Q_f - Q_i = \text{net signed area bounded by the rate-of-change curve} \tag{145}$$
$$\text{and the } t\text{-axis, from } t_i \text{ to } t_f.$$

Here's how we'll think about the derivation. Imagine that
you are driving steadily along a straight, level road, at a speed
of $v_{i,x}$. Then you begin to speed up at a constant rate a_x over a
period of T seconds. This brings your speed up to $v_{f,x}$.

a. Sketch labeled $a_x - t$, $v_x - t$, and $x - t$ graphs for the motion. Show a range of times starting before the acceleration
   phase begins and ending after the acceleration phase ends.

b. Use Equation 145 to show that $\Delta v_x = a_x T$. Label your $v_x - t$
   graph with this information.

c. Use your $v_x - t$ graph along with Equation 145 to show that
   $\Delta x = v_{i,x}T + \frac{1}{2}a_x T^2$. You have just derived Equation 144.

14. [Uses calculus.] Starting from the mathematical definition of
velocity, $v_x = \frac{d}{dt}x$, apply the Fundamental Theorem of Calculus
to conclude that

$$\Delta x = \int_{t_i}^{t_f} v_x(t)\, dt. \tag{146}$$

You have just derived Equation 84.

15. [Uses calculus.] (This is a follow-up from Focused Problem 5 of this chapter; you might want to do that problem first.) A taxi driver wants to drive from a stop sign at $x = 0$ to the next stop sign at $x = D$. He accomplishes this with a velocity profile $v_x(t) = \frac{8v_0^2}{3D}\left(t - \frac{2v_0}{3D}t^2\right)$, where v_0 is the local speed limit.

a. Show that with this velocity profile the cab starts from rest, reaches a maximum speed of v_0 at time $t_{max} = \frac{3D}{4v_0}$, and then comes to rest again at time $t_{stop} = \frac{3D}{2v_0}$.

b. Show that the cab travels a distance D by the time it comes to rest.

c. Find $a_x(t)$ and $x(t)$ as explicit functions of time. Choose reasonable values for v_0 and D, and graph $x(t)$ and $a_x(t)$ on a graphing calculator. Do your graphs resemble the $x - t$ and $a_x - t$ graphs you drew in Focused Problem 5?

SOLUTIONS CAN BE FOUND ON PAGE 372

PART II EXPLAINING AND PREDICTING MOTION

Using the vectorial concepts of position, velocity, and acceleration, we can now describe motion in exquisite detail. But why does motion happen in the first place? What forces in the universe cause things to change? How exactly do these forces animate the universe?

CHAPTER 10

The Concept of Force

In Chapter 6 we saw that in physics the word *acceleration* means much more than simply "speeding up," its meaning in everyday speech. Physics also uses other words from everyday speech: words like *force, momentum,* and *energy,* for example. In casual speech, *force, momentum,* and *energy* can be more or less synonymous. But to a physicist, these words name entirely different things.

This chapter has no formulas, but it's one of the most important in this book!

In this book we won't be going into detail about momentum or energy. We're concerned with force alone. But for the sake of contrasting force with momentum and energy, let's just say a few words about all three of these concepts.

Start with energy. We all know that a person can be "energetic." And we often say, for instance, that "children have a lot of energy." Apparently energy is the kind of thing you can *have*—like money, say. As I like to put it, energy is "havable." In this respect, everyday language matches up well with physics. An eight-year-old bouncing around a room "has a lot of energy" not only in the colloquial sense but also in the physics sense of having "kinetic energy," or energy of motion.

When we observe the eight-year-old bouncing around the room, our feeling that she is energetic has nothing to do with the *direction* in which she is moving. We would call her energetic whether she were moving left or right at any given moment. It is really only the speed factor that we're paying attention to. This also matches up well with physics, because in physics kinetic energy is

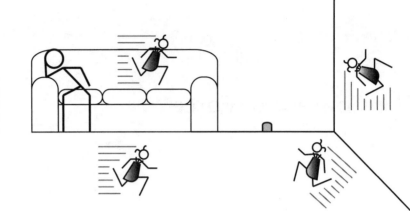

FIGURE 158.
Little Jane running
around a room. She
has a lot of kinetic
energy.

a scalar quantity, not a vector quantity; the direction of motion is irrelevant. See Figure 158.

What about momentum? We often hear athletes saying things like this: "By the third quarter we had the momentum, and we just coasted to the end." We *had* the momentum; apparently, momentum is another havable thing, something that, like money, you can have, gain, or lose. And the bit about "coasting to the end" also suggests, rightly, that momentum has to do with *motion*. If you're moving, you have momentum. If you're moving really fast, you have more momentum. If you're moving really fast and you're a really big linebacker, you have even more momentum! I like to say that momentum is a kind of "oomph" factor. It's the amount of oomph that a moving object possesses. You don't want to be tackled by somebody with a lot of oomph.

So far, the everyday meaning of *momentum* is matching up well with the physics meaning. The only nuance, perhaps, is that in physics momentum always has a direction, or arrow, associated with it. Momentum is one of those vector quantities. Naturally enough, the direction of your momentum vector at any given time is simply the direction in which you are moving at that instant. See Figure 159. The two rams in the figure move with equal speeds. But the big ram has more momentum, or oomph.

You may be curious about how we calculate the amount of oomph contained in a moving object. Because the rule is so easy, I'll just state it here. The momentum possessed by a moving object is equal to the object's mass, *m*, times its velocity vector, **v**. We usually write the rule as

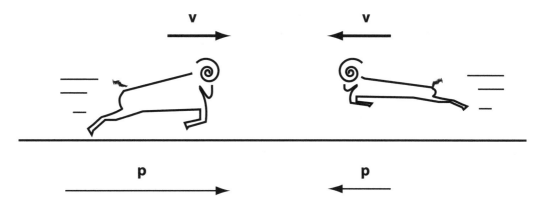

FIGURE 159.
The velocity vectors (**v**) and momentum vectors (**p**) for two rams.

$\mathbf{p} = m\mathbf{v}$, where $\mathbf{p}$ denotes the momentum vector of the moving object. Momentum is well represented by an arrow, because you carry your oomph in a particular direction.

OK, now for the formula for kinetic energy, in case you're interested. The kinetic energy possessed by a moving object is equal to half the object's mass times the square of its speed. We usually write this rule as $K = \frac{1}{2}mv^2$, where K denotes the kinetic energy. Note that there's no boldfaced variable anywhere in this rule; kinetic energy is a scalar quantity with no sense of direction to it.

This brings us to our main concern, the concept of force.

The word *force* is used in many different ways in everyday speech. Only some of these uses match up well with the physics concept of force. We'll get to some instructive examples, but without beating around the bush, let me tell you what a force is in physics. In the most general sense,

A force is an influence of one object on another.   (147)

I repeat: A force is an influence of one object on another.

As we've seen, sometimes everyday language matches up well with the physics concept. But not always. For example, consider the following snippet of conversation, overheard in the lobby of the Holiday Inn in Des Moines:

Alice: You can't force me to love you, Bob.
Bob: [Whimpers.]

Bob would very much like to *influence* Alice. He would like to *change* what Alice would naturally do on her own. If Bob *could* "force" Alice to love him, this would indeed be an example of one object influencing another.

Force Is Not Havable

Let me restate our definition: A force is an influence of one object on another object. From this it should be apparent that force is not something one object can have by itself. Force is not havable.

Although you can't *have* force, you can *exert* force (on something else). This is how we make sense of a word like *forceful.* If we say that Teddy Roosevelt was a forceful person, we can't very well mean that he possessed a lot of force. That would be nonsense. What we must mean is that Teddy Roosevelt was good at exerting strong forces on other people.

Every Force Has a Source Object and a Target Object

In physics, every force has a *source* object and a *target* object. For example:

"I push the wall."
"The ground pushes on my foot."
"The earth pulls downward on the cannonball."

In each sentence, the source object is single-underlined, and the target object is double-underlined.

» WORKED EXAMPLE 1

Following are two sentences from everyday language. For each sentence, explain how the use of the word *force* differs from the physics usage, and also rewrite the sentence so that it uses *force* in something resembling the correct physics sense.

I helped her out of a jam, I guess,
But I used a little too much force.

—Bob Dylan, from "Tangled Up in Blue"

Eakins is not a painter, he is a force.

> —Walt Whitman, commenting on his portrait by Thomas Eakins

» SOLUTION 1

The problem with Dylan's use of the word *force* is that he makes it sound as though force is a substance that can be doled out—you can use too much, you can use too little—like garlic. ("I got the pasta done, on time, / But I used a little too much garlic.") To rescue the usage, we might say something like this: "I helped her out of a jam, I guess, / But (in doing so) I exerted too much force on her." (Hardly works as poetry, does it?)

Whitman's use of *force* is a complete mess. Because a force is an influence of one object on another object, a lone person can't *be* a force. Perhaps we could rescue his expression by saying, "Eakins is not just a painter, He exerts a tremendous force on all of us." (Ugh.) ◇

I have defined a force as an influence of one object on another. Many textbooks define a force as a push or pull. To the extent that *push* and *pull* are synonyms for *influence* (as in, "I felt Dracula's influence pulling me toward him"), I like the push/pull definition. However, when a textbook simply says, "A force is a push or pull," it is leaving out something very important: the fact that there must always be a push*er* (the source) and a push*ee* (the target).

Forces Always Come in Pairs

The force relationship between two objects is always mutual. For example, suppose Alice and Bob are leaning against each other, as shown in Figure 160.

What stops Bob from falling over? It is the force that Alice exerts on him. And what stops Alice from falling over? It is the force

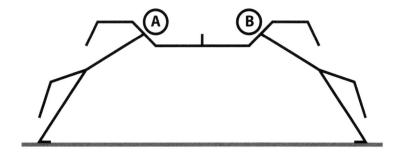

FIGURE 160.
Alice and Bob leaning against each other.

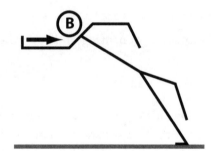

FIGURE 161.
Bob as target.

that Bob exerts on her. Whenever A exerts a force on B, B also exerts a force on A. You can't touch without being touched.

Here's another example of how forces always come in pairs. Imagine standing barefoot on your front lawn. In this situation, you will exert a steady downward force on the grass. The evidence of this is found in the footprints you will leave in the soft turf. But at the same time, the grass is also exerting an upward force on you. The evidence for this is found in the threadlike impressions that will appear on the soles of your feet. How else were these tiny dents formed but by the action of individual blades of grass pressing against your skin?

Force goes both ways. Whenever you press on the grass, the grass presses back on you. If I shake hands with you, equally well you're shaking hands with me. Again, you can't touch without being touched.

A Force Diagram Focuses on a Single Target

In Figure 160, Alice probably thinks *she's* doing the pushing. Bob would probably say the same thing. In truth, both of them are right. Alice pushes on Bob, and Bob pushes on Alice. To analyze this situation any further, you will have to make a choice and arbitrarily decide whom you will regard as the target of the force. For example, suppose we agree to consider Bob as the target. Then we can draw a picture like that in Figure 161. I have drawn the force acting on Bob as an arrow to the right, because Alice is pushing Bob to the right.

Of course, we might just as well have decided to focus on Alice as the target. Then we would have drawn a picture like that in Figure 162. I have drawn the force acting on Alice as an arrow to the left, because Bob is pushing Alice to the left.

Notice that Alice is nowhere to be seen in Figure 161. When drawing a diagram of the target of force, keep the focus on the chosen target; don't show any other objects in the diagram.

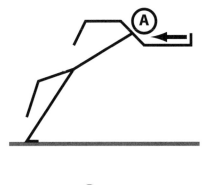

FIGURE 162.
Alice as target.

FIGURE 163.
Patting yourself on the back. Go ahead. You deserve it.

The two forces shown in Figures 161 and 162 are *different forces.* Both forces exist during the time when the two people are leaning against each other. You can see that the two forces are different because the two vectors are different; one vector points left, while the other vector points right.

Notice that Bob is nowhere to be seen in Figure 162.

Objects Can Be Subdivided

Can there ever be a situation in which the source object and the target object are actually the same? For example, what happens when you pat yourself on the back (as in Figure 163)? Aren't you both the source and the target here?

One could argue that the answer is yes, but a more illuminating way to analyze a situation like this is to proceed by mentally dividing your body into two parts. For example, you might consider your arm as one object and consider your body as another object. Now we can proceed by, say, considering your body as the

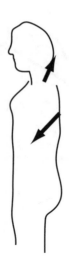

FIGURE 164.
Your body as the target. The forces exerted on your body by your arm when you pat yourself on the back.

FIGURE 165.
Your arm as the target. The forces exerted on your arm by your body when you pat yourself on the back.

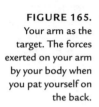

target. Figure 164 shows the forces exerted on your body by your arm. Your hand presses against your back, and your upper arm tugs at your shoulder socket. Only your body is shown.

Alternatively, we could consider your arm as the target. Figure 165 shows the forces exerted on your arm by your body. Your back presses against your hand, and your shoulder socket tugs on your upper arm. Only your arm is shown.

Does it surprise you to think that your body exerts a force on your hand when you pat yourself on the back? Try it with your eyes closed. Ask yourself: Can I feel anything with my fingers while this is happening?

Of course you can. What your fingers are feeling is the force exerted on them by your back. To your fingers it feels much the same as if someone were tapping your fingertips with their own.

This trick of mentally dividing an object into parts is frequently useful in force analysis. Mechanical engineers mentally divide bridges and buildings into many different parts, considering some of the subparts as sources and other subparts as targets. The famous thought-experiment of Galileo Galilei—the one questioning Aristotle's assertion that heavier objects fall faster than light objects—also relies on a clever mental division of an object into parts.

Even Inanimate Objects Push Back

When you push a shopping cart down the aisle, it is quite natural to say that you are exerting a force on the cart. It may seem less natural to say that the cart is exerting a force on you. After all, who's in charge here? *You're* the one doing the pushing, right? And yet we have said that force is always a two-way street. Forces always come in pairs. Therefore, if you are pushing the cart, the cart is also pushing back on you. Indeed, think about what would happen if the handle of the shopping cart were covered with sharp tacks. In this case I think you would agree that your hands would register an influence coming from the shopping cart—a painful influence, in fact!

In everyday speech, pushing and pulling are active verbs that presuppose a conscious agent who pushes or pulls because of some intention. But pushing and pulling in the physics sense have nothing to do with who is driving the action, who is in command, or who wills events to happen. Those are human considerations, not physical ones. In our universe, even inanimate objects can exert their influence on other things.

The Strength of a Force Is Measured in Newtons

An influence can be strong or weak. A push or a pull can be strong or weak. When you exert a force on something, the force can be strong or weak.

The strength of a force is indicated by the magnitude of the force vector used to represent it. For example, if you look at Figures 161 and 162, it appears that the force exerted on Bob by Alice is equal in strength to the force exerted on Alice by Bob. (At least that's what is indicated in the diagram; we leave until Chapter 15 the question of whether this must be the case in reality.)

The strength of a force is measured in units called *newtons* (N), in honor of Sir Isaac Newton. A 1-N force is a pretty weak force in human terms. If you pat yourself on the back, your hand is probably exerting a force of about 1 N on your back. (More on this in Chapter 13.)

Forces Can Turn On and Off

Forces are evanescent things. They're not like material objects. They appear and disappear all the time. When you and I are shaking hands with each other, you're exerting a force on me, and I'm exerting a force on you. Both of these forces vary with time as our clasped hands pump up and down, and once we let go of each other's hands, both of these forces simply vanish. The vectors disappear and are no more.

A Force Diagram Illustrates a Single Instant in Time

Because forces come and go, varying in strength from moment to moment, a force diagram is generally meant to illustrate only a *single moment of time*. Whenever you draw a force diagram, make sure you're not confused about which moment in time you're diagramming.

Force Requires Contact

Suppose I'm standing at the front of the classroom and I tell you to stand up. You then stand up. I certainly influenced you, didn't I? I mean, you wouldn't have stood up if I hadn't told you to, right?

Though it may seem strange to view a social interaction like this as a physical process, our little encounter certainly was a physical event. One animal uttered a bark of sorts, and another animal reared up on its hind legs in response. The anthropologist from Mars jotted it down in a field notebook and later published it in *Proceedings of the Martian Academy of Sciences*. Now it's a recorded scientific fact.

So here's my question: *How did this influence carry across space from me to you?*

Well, one part of the story goes something like this. My vocal cords started vibrating. And, just as if they were guitar strings, these vibrating cords bumped the neighboring air molecules into rhythmic motion. These molecules bumped into their neighboring molecules, and they into their neighbors, and so on, until the molecules right next to your eardrums were bumped into motion, "in tune" with the music of my vocal cords. These molecules then bumped into your eardrum, setting *it* into motion in time with the music. Next, through some sort of electromechanical coupling, the motion of your eardrum created electrical currents in your nerves, and (skipping the last and least understood part of the story) you decided to stand up.

The point is that in order for this influence to travel from me to you, there had to be a chain of *contact* between various material objects: my vocal cords knocking air molecules, air molecules knocking other air molecules, and those other air molecules knocking against your eardrums.

Evidently, then, when Harry Potter uses a *Wingardium Leviosa* charm to lift a car into the air, something funny is going on. Because as far as I can tell from the movies I've seen, the influence appears to leap across empty space from Harry to the car. Such a thing can't happen. Influences need a chain of physical contact in order to travel from one object to another.

Another thing that marks this fictional encounter as make-believe is that Harry appears to exert an influence on the car without the car's exerting any noticeable influence on him. This is action without *inter*action. That can't happen, because forces always come in pairs. Newtonian mechanics says that if Harry isn't strong enough to lift a car with his arms, he won't be strong enough to lift it with a wand, either. (I know, I know; that's why they call it "magic.")

When I said that force can be transmitted only by physical contact between objects, you may have started worrying about gravity. After all, the sun seems to pull on the earth across empty space, doesn't it? And when you toss an apple up into the air, doesn't the earth pull the apple back down without touching it? After all, it's not as if the earth reaches up a stony arm to grab the apple! So isn't the apple an example of influence without contact?

If you're worried about this, you should be. Newton himself certainly was. The best theory of gravity that Newton could come up with involved "magical" action at a distance. In Newton's *Principia*, gravitational influences mysteriously connect source and target across empty space. Newton himself knew this was a ridiculous idea. It was just that (1) he couldn't think of a better theory and (2) the laws that he did put forth appeared to account for the observed motions of the planets. So in his scientific writings, Newton let the matter rest. But he shared his true thoughts on the subject in a private letter to the Reverend Richard Bentley in 1693:

> That gravity should be innate, inherent and essential to matter, so that one body may act upon another at a distance through

a vacuum, without the mediation of any thing else, by and through which their action and force may be conveyed from one to another, is to me so great an absurdity, that I believe no man who has in philosophical matters a competent faculty of thinking, can ever fall into it [4].

Today we know how gravitational influence travels from one object to another. Or at least we know how to tell a pretty good story about how it happens. The story is the theory of general relativity, which was developed by Albert Einstein in 1915 and has been well supported by a large body of detailed experimental evidence gathered since then. General relativity describes how gravitational influence travels from one object to another via ripples in the fabric of spacetime. In fact, as I write this page, devices are actually being built that may, for the first time, detect gravitational influences in midflight. Newton would have been thrilled!

In this book, though, we'll be treating gravity the way Newton originally did. We'll assume that the sun and planets somehow exert forces on each other across empty space, just as do the earth and a bird in flight. Other examples of "magical" situations would be when an electron and a proton exert forces on each other across the empty space in an atom or when a magnet pulls a nail across a tabletop without touching the nail.

However, *in general,* try to remember this: If two things aren't touching, they can't be exerting forces on each other.

Deciding What Forces Are Acting on a Chosen Target

There is no formula that will tell you what forces are acting on a chosen target at any given moment. The best tool you have for recognizing the forces that act on something is your own physical intuition about the world. This intuition will continue to sharpen as you study physics.

»WORKED EXAMPLE 2

A push broom is leaning against a wall, as shown in Figure 166. What forces are acting on the broom at this moment? Make a diagram showing these forces as vectors.

»SOLUTION 2

A good way to bring your intuition to bear on this situation is to imagine that *you* are the broom. If you were leaning against a wall

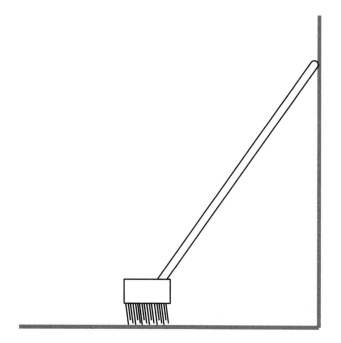

FIGURE 166.
Worked Example 2. A push broom leaning against a wall.

like that, what would you feel? Try leaning against a wall with your shoulder blades against the wall. (And take your shoes off so you can feel everything.) Close your eyes, and search your senses. What do you feel?

You feel pressure in your shoulders where they touch the wall, perhaps even some pricking if the wall is rough and the skin of your shoulders is exposed. Evidently, the wall is pushing on your shoulders.

Down at the floor, you feel compression in the soles of your feet where they touch the floor. Evidently, the floor is pushing upward against the soles of your feet, compressing your flesh.

You also feel the skin on the bottoms of your feet tending to roll back, making the skin around your toes feel tight. Evidently, the floor is dragging the soles of your feet backward toward the wall. (And did you look closely at the bristles of the push broom in Figure 166? Do you see the way the floor is bending the bristles towards the wall? What else is bending these bristles, if not the floor?)

Putting these insights together, I can draw a diagram like that shown in Figure 167.

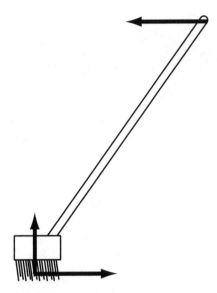

FIGURE 167.
Partial solution to Worked Example 2. A force diagram for the broom in Figure 166.

An equally good way to draw the diagram would be as in Figure 168. This figure combines the pressing-up and pulling-back actions of the floor into one force: the composite, or *net* force, exerted on the broom by the floor. (We'll see more about manipulations like this in Chapter 11.)

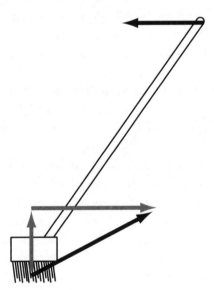

FIGURE 168.
Worked Example 2. A force diagram for the broom that is equivalent to the one shown in Figure 167.

Finally, as you might already have realized, there is one other force that belongs in the diagram: the gravitational force that the earth exerts on the broom. This is not something one can easily feel by introspection; we live our entire lives in a gravitational field, and we lack the ability to sense gravitation per se. To know that you should include a gravitational force acting on the broom, you just have to know enough about the world we live in to know that the earth's gravity pulls downward on everything all the time. We'll learn more about gravity in Chapter 16, but for now, Figure 169 shows how we'd include the gravitational force on the broom in the diagram. The earth pulls downward on the broom (with its magical hand that reaches across empty space).

For the purposes of this book, it doesn't much matter where you "attach" a gravitational force to its target. Properly speaking, you should attach a gravitational force to the target's *center of gravity point*. But finding the center of gravity point is outside the scope of this book, and it's not something we need to worry about. None of the answers in this book depends on the specific point of attachment you choose.

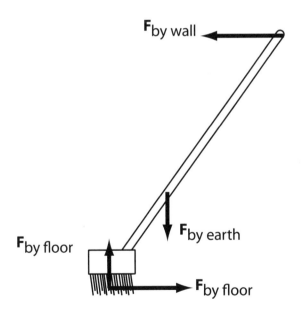

FIGURE 169.
All of the important forces on the broom in Worked Example 2. The object exerting each force has been identified with subscripts.

You may be wondering if there are any other forces acting on the broom. There certainly are! The moon is not all that far away—only 250,000 miles or so. This is not exactly a stupendous distance; it's more like a healthy balance in your frequent-flier account. The moon exerts a gravitational force on the broom, just as the earth does. But the gravitational force exerted on the broom by the moon is only about 0.001 N. This is small compared to the earth's gravitational force acting on the broom (about 5 N, say). Still, if you wanted to be very, *very* precise about everything, you'd have to take the lunar gravitational force into account.

Actually, even larger than the force exerted by the moon is the force exerted by the sun. The sun is much farther away, but it's also much more massive, so its gravitational field is stronger than the moon's gravitational field in the vicinity of the earth. The force exerted on the broom by the sun is about twice as strong as the force exerted on the broom by the moon!

It's always up to you to decide what forces are sufficiently strong to be included in a force diagram. For example, what if there were a strong breeze flowing past the broom? The force exerted by the moving air might be strong enough to knock the broom over. To find out what would happen, we'd have to include the wind force in the diagram. ◇

» WORKED EXAMPLE 3

A hunter fires a bullet across an open field, and the bullet embeds itself in a tree. Draw a force diagram showing the forces that act on the bullet at a moment in time when the bullet is halfway across the field. For simplicity, neglect the force of air resistance on the bullet.

» SOLUTION 3

Remember that a force diagram always illustrates a *single moment of time*. The moment of time we are considering is when the bullet is halfway across the field. My diagram for this moment is shown in Figure 170.

FIGURE 170.
Solution to Worked Example 3. Force diagram for a bullet at a moment in time when it is halfway across a field.

$F_{\text{by earth}}$

There is only one force in my diagram. That's because I can think of only one *object* that is exerting an influence on the bullet *at this moment:* the earth. (Remember that for simplicity we are neglecting the force exerted on the bullet by the air.)

Almost everyone draws this force diagram incorrectly at first. Almost invariably, the first-time physics student will draw the diagram as shown in Figure 171.

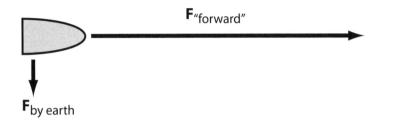

FIGURE 171.
An incorrect force diagram for the moment of time described in Worked Example 3.

When I see a student draw a diagram like this, with an extra "forward" force in it, our conversation usually runs something like this:

> *Me:* Hmm. . . . My diagram has only one force in it, but yours has two. Why did you put the extra "forward" force in there?
>
> *Bob:* The forward force is the force that keeps the bullet going across the field.
>
> *Me:* Fair enough. But remember that a force is an influence of one object on another object. So, what *other object* would you identify as being the source of the "forward" force?
>
> *Bob:* Well, the rifle.

You see what's wrong with Bob's answer, don't you? *The rifle is no longer touching the bullet.* Force requires contact. If the rifle isn't touching the bullet anymore, it can't very well be exerting a force on the bullet anymore. I would agree that *when the bullet was in the rifle's barrel,* the rifle was exerting huge forces on the bullet (and vice versa). But as soon as the bullet emerged from the muzzle, that handshake was over, so to speak, and those forces *vanished.*

Returning to our dramatic scene, already in progress, Bob has just been reminded that the rifle is no longer touching the bullet and therefore cannot be exerting a force on it:

> *Bob:* But if there's no forward force, what keeps the bullet moving across the field?

Now we reach the heart of the matter. Bob's instinct is that *something* must keep the bullet moving across the field. I'd say that's a perfectly reasonable instinct. But Bob's mistake is to seize on *force* as the sort of thing that keeps the bullet going. This simply can't be right, because no *other object* is touching the bullet in such a way as to push it forward. Clearly, then, the "something" that keeps the bullet moving across the field is not a force at all. Rather, the bullet must have some kind of *internal* "desire" to keep moving once it has been set in motion by the rifle. ◇

Newton termed this internal desire *inertia*. We'll be looking at inertia in more detail later (in Chapters 12 and 13). But for now, let me end Chapter 10 with just one final observation on the concept of force in physics.

Force Is a Vector Quantity

We have been using arrows (vectors) to represent forces in our diagrams. Every force has a magnitude, or strength, measured in newtons, and every force has a direction in space: the direction of the push or pull.

Some of the lighthearted examples of forces in this chapter were *not* vectorial in nature, which means they were not really forces in the physics sense. When Bob tried to "force" Alice to love him, this was not really a physics example of force! If a force is a physics force, it must have a quantifiable strength in newtons and an identifiable direction in space.

Reader's Notes

Below is space for you to summarize key points from this chapter.

FOCUSED PROBLEMS :: CHAPTER 10

1. A little box rests on a big box, which rests on the ground (Figure 172). Draw a force diagram for the little box and a separate force diagram for the big box. For each force vector appearing in your diagrams, use a subscript "by . . . " to identify the source object of the force.

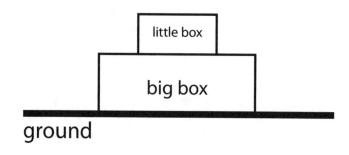

FIGURE 172.
Focused Problem 1.
A little box resting on
a big box resting
on the ground.

2. An orangutan hangs from a monkey bar (Figure 173). Draw a force diagram for the orangutan and a separate force diagram for the monkey bar. For each force vector appearing in your diagrams, use a subscript "by . . . " to identify the source object of the force.

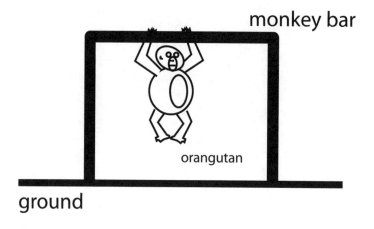

FIGURE 173.
Focused Problem 2.
An orangutan
hanging from a
monkey bar.

3. Figure 174 shows a highly simplified diagram of a person standing on tiptoe. Draw a force diagram for the *foot bone only*, showing all of the significant forces acting on the foot bone. (Yes, I know there are more bones than this in each foot

and leg, but for the sake of a simple model we will pretend that the anatomy is as shown in the figure.)

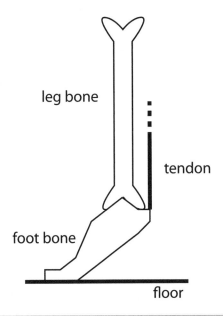

FIGURE 174.
Focused Problem 3. Simplified diagram of the foot and leg bones of a person standing on tiptoe.

SOLUTIONS CAN BE FOUND ON PAGE 387

CHAPTER 11

Combining Forces That Act on the Same Target

At any given time, your body is subject to a number of noticeable influences. The earth pulls you down. Your chair pushes you up. The moon and sun tug on you slightly, the other celestial bodies negligibly so. Now a breeze ruffles your hair; the air is exerting a force on you as well. All of these forces can be combined, leading to a single net influence. This net influence is called the *net force* acting on you at any given time.

Finding the Net Force on a Target

Here is how to find the net force on a target:

1. Draw a force diagram for the target, showing all of the significant force vectors acting on the target.
2. Add all of the force vectors together. The result of your vector addition is a single vector, the net force. We denote the net force by $\mathbf{F}_{net}$.

As we shall see in more detail in the next chapter, it is the *net* force that actually governs the motion of an object.

Our friends Alice and Bob from Chapter 9 have decided a long-distance relationship doesn't work for them. They've moved to Nashville and found jobs at the post office. At a company picnic, they lead opposing teams in a tug-of-war.

Let's see how net force comes into play in the tug-of-war. Team Bob might be pulling on the rope with a force of 200 N to the right,

Now would be a great time to revisit Chapter 4. You really won't be able to survive physics unless you can add vectors fluently.

but that doesn't necessarily mean the rope will move to the right! After all, what if Team Alice is pulling with a force of 250 N to the left? You have to know all of the significant forces acting on a target in order to decide what will happen to the target.

» WORKED EXAMPLE 1

At a crucial moment in the tug-of-war, Team Alice is pulling on the rope with a force of 250 N to the left, and Team Bob is pulling on the rope with a force of 200 N to the right. All other forces acting on the rope are negligible. Find the net force acting on the rope.

» SOLUTION 1

Figure 175 shows a cartoon of the situation. Notice that I have included a coordinate system in my cartoon; you should always do this when diagramming a quantitative problem.

FIGURE 175.
Worked Example 1.
Team Alice and Team
Bob in a tug-of-war.

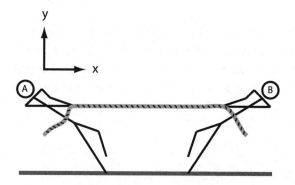

Figure 176 isolates the rope as our target and shows the two forces acting on the rope. The *rope* is the target here; Alice and Bob are not even shown in the diagram. I always like to put the tail of a force vector right at the point where the force is being applied to the target. By the way, I am neglecting the force exerted on the rope by the earth.

FIGURE 176.
Solving Worked
Example 1. A force
diagram for the rope
in Figure 175.

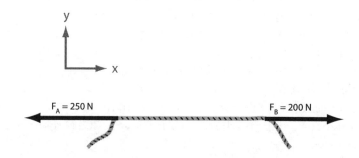

$F_A = 250\,\text{N}$ $F_B = 200\,\text{N}$

The net force on the rope is found by adding the two vectors $\mathbf{F}_A$ and $\mathbf{F}_B$. To begin with, I always like to do the addition graphically, by the tip-to-tail method. That way, I can get a sense of what my final answer is supposed to look like. Adding the forces tip-to-tail gives a net force as shown in Figure 177.

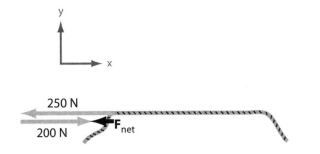

FIGURE 177.
Solution to Worked Example 1. Graphical analysis showing the net force vector $\mathbf{F}_{net}$ acting on the rope in Figure 175.

So the net force is rather small in magnitude and points to the left. A quantitative approach is as follows:

$$F_{net,x} = F_{A,x} + F_{B,x} \tag{148}$$

$$= (-250\,\text{N}) + (+200\,\text{N}) \tag{149}$$

$$= -50\,\text{N}. \tag{150}$$

$$F_{net,y} = F_{A,y} + F_{B,y} \tag{151}$$

$$= (0\,\text{N}) + (0\,\text{N}) \tag{152}$$

$$= 0\,\text{N}. \tag{153}$$

The final answer for $\mathbf{F}_{net}$ can be presented in various ways. We could simply present the x- and y-components of $\mathbf{F}_{net}$ as follows:

$$F_{net,x} = -50\,\text{N} \tag{154}$$

$$F_{net,y} = 0\,\text{N}. \tag{155}$$

We could also present these components in the form

$$\mathbf{F}_{net} = (-50\,\text{N}, 0\,\text{N}). \tag{156}$$

Alternatively, we could specify the magnitude and direction of $\mathbf{F}_{net}$:

$$F_{net} = \sqrt{F_{net,x}^2 + F_{net,y}^2} \tag{157}$$

$$= \sqrt{(-50\,\text{N})^2 + 0^2} \tag{158}$$

$$= 50\,\text{N}. \tag{159}$$

The net force points directly to the left, with magnitude 50 N. (This was already pretty obvious from our tip-to-tail work in Figure 177.)

◇

» WORKED EXAMPLE 2

At work, Alice and Bob are pushing a mail cart along a horizontal floor. Each pushes with a force of 10 N, but they push at different angles, as shown in Figure 178. All other forces acting on the cart are either negligibly small or not important for the analysis we're going to do.

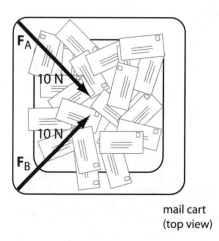

FIGURE 178.
Worked Example 2. Alice and Bob pushing on a mail cart, each with a force of 10 N.

mail cart
(top view)

Without doing any detailed calculations, what do you think is the direction of $\mathbf{F}_{net}$ at the moment of time shown? And without doing any detailed calculations, do you think F_{net} (the magnitude of the net force) is greater than, equal to, or less than 20 N? Finally, find $\mathbf{F}_{net}$ quantitatively.

» SOLUTION 2

Let's start with the intuitive questions. First of all, think about directions. Alice is pushing down and to the right, while Bob is pushing up and to the right. (Here I mean "up" and "down" with reference to the top and bottom of the page, not with reference to Alice's and Bob's own sense of up and down.)

To the extent that Alice is pushing partly down and Bob is pushing partly up, their efforts partly offset one another. Alice's down cancels Bob's up, leaving only the joint rightward push to propel the cart. In other words, $\mathbf{F}_{net}$ points to the right.

Now for the magnitude. Alice is pushing with a strength of 10 N, and Bob is pushing with a strength of 10 N as well. So it might seem natural to conclude that the net force acting on the cart is 20 N. But this ignores what we just decided earlier: that Alice's effort and Bob's effort partially cancel each other. The downward part of Alice's push is canceled by the upward part of Bob's. Only the rightward part of Alice's push contributes to the net effect, and this rightward part is less than entire amount of 10 N. The same goes for Bob's part of the push. So in finding the net force, we'll be adding two numbers less than 10 N, which means that our sum will be less than 20 N in strength. In short, we have two forces of 10 N that partly work against each other, so we'll find that the net force is less than 20 N in strength. If Alice and Bob want to join forces (heh, heh) and push with a net force of 20 N, they should move together on the same side of the cart and push in the same direction!

Before we find $\mathbf{F}_{net}$ quantitatively, I like to find $\mathbf{F}_{net}$ graphically by the tip-to-tail method. Figure 179 shows the net force obtained in this way.

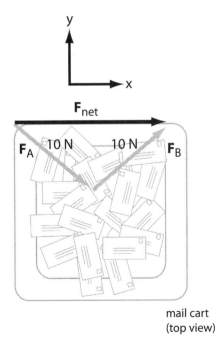

mail cart
(top view)

FIGURE 179.
Graphic solution to Worked Example 2. Finding the net force graphically using the tip-to-tail method.

Now for a quantitative approach. To add the vectors $\mathbf{F}_A$ and $\mathbf{F}_B$, we have to break both vectors into components, then add component-wise.

Figure 180 shows the forces broken into components. I used trigonometry to do this. Notice that the original vectors $\mathbf{F}_A$ and $\mathbf{F}_B$ are not in the diagram anymore. This is because $\mathbf{F}_A$ and $\mathbf{F}_B$ are entirely equivalent to the sum of their individual components. So I have replaced $\mathbf{F}_A$ and $\mathbf{F}_B$ with their equivalents.

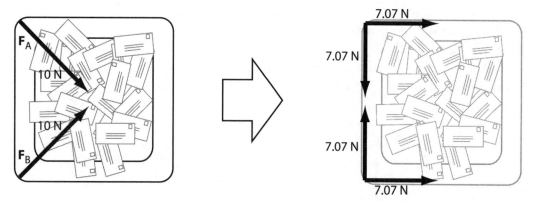

FIGURE 180. Worked Example 2. The original forces broken into components.

With reference to Figure 180, we find the components of the net force vector as follows:

$$(160) \qquad F_{\text{net},x} = F_{A,x} + F_{B,x}$$

$$(161) \qquad = (+7.07\,\text{N}) + (+7.07\,\text{N})$$

$$(162) \qquad = 14.1\,\text{N}.$$

$$(163) \qquad F_{\text{net},y} = F_{A,y} + F_{B,y}$$

$$(164) \qquad = (-7.07\,\text{N}) + (+7.07\,\text{N})$$

$$(165) \qquad = 0\,\text{N}.$$

The final answer for $\mathbf{F}_{net}$ can be presented in various ways. We could simply present the x- and y-components of $\mathbf{F}_{net}$ as follows:

$$F_{net,x} = +14.1\,\text{N} \tag{166}$$

$$F_{net,y} = 0\,\text{N}. \tag{167}$$

We could also present these components in the form

$$\mathbf{F}_{net} = (+14.1\,\text{N}, 0\,\text{N}). \tag{168}$$

Alternatively, we could specify the magnitude and direction of $\mathbf{F}_{net}$:

$$F_{net} = \sqrt{F_{net,x}^2 + F_{net,y}^2} \tag{169}$$

$$= \sqrt{(+14.1\,\text{N})^2 + 0^2} \tag{170}$$

$$= 14.1\,\text{N}. \tag{171}$$

The net force points directly to the right, with magnitude 14.1 N. By way of checking our answer, the components $F_{net,x} = 14.1\,\text{N}$ and $F_{net,y} = 0\,\text{N}$ do, in fact, look like our graphical solution in Figure 179. So everything looks good. ◇

» WORKED EXAMPLE 3

Alice and Bob are trying to push a mail cart into Cecil's office, but Cecil doesn't want the mail cart because he'll have to sort all of that mail. So Cecil is pushing back. The forces are shown in Figure 181.

The forces applied by Alice and Bob are given. What force $\mathbf{F}_C$ must Cecil apply in order to cancel out the opposing forces exactly and create a situation with zero net force on the cart?

» SOLUTION 3

I can think of two good approaches to this problem. We'll see how both of them work out.

One approach would be to consider the x- and y-components of Cecil's force as unknowns, then solve for them by requiring that both components of the net force vanish: $F_{A,x} + F_{B,x} + F_{C,x} = 0$, and $F_{A,y} + F_{B,y} + F_{C,y} = 0$.

For this we need the x- and y-components of $\mathbf{F}_A$ and $\mathbf{F}_B$. These are shown in Figure 182 based on the trigonometry of the given situation.

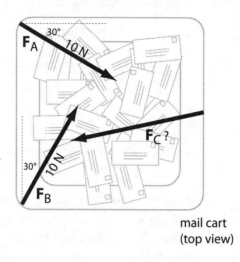

FIGURE 181.
Worked Example 3.
Alice, Bob, and
Cecil pushing on
a mail cart.

mail cart
(top view)

Figure 182 shows Cecil's force $\mathbf{F}_C$ broken into components as well, though we don't have numerical values for these components yet.

In terms of quantities in the diagram, the x-component of the net force vector is given as follows:

$$F_{\text{net},x} = F_{A,x} + F_{B,x} + F_{C,x} \tag{172}$$

$$= (+8.66\,\text{N}) + (+5\,\text{N}) + F_{C,x} \tag{173}$$

$$F_{\text{net},x} = 13.66\,\text{N} + F_{C,x}. \tag{174}$$

In order for $F_{\text{net},x}$ to vanish, we need $F_{C,x} = -13.66\,\text{N}$. We expected a negative answer for the x-component because Cecil must be basically pushing to the left (on the page), not to the right.

Now let's do the same process for the y-component of the net force:

$$F_{\text{net},y} = F_{A,y} + F_{B,y} + F_{C,y} \tag{175}$$

$$= (-5\,\text{N}) + (+8.66\,\text{N}) \tag{176}$$

$$F_{\text{net},y} = 3.66\,\text{N} + F_{C,y}. \tag{177}$$

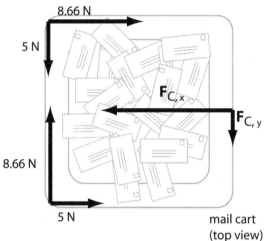

FIGURE 182.
Solving Worked
Example 3. The
original forces shown
in Figure 181 broken
into components.

In order for $F_{net,y}$ to vanish, we need $F_{C,y} = -3.66$ N. Apparently Cecil will have to push slightly downward (on the page) to cancel Alice and Bob's forces.

The final answer for $\mathbf{F}_{net}$ can be presented in various ways. We could simply present the x- and y-components of $\mathbf{F}_C$:

$$F_{C,x} = -13.66 \text{ N} \tag{178}$$

$$F_{C,y} = -3.66 \text{ N}. \tag{179}$$

We could also present these components in the form

$$\mathbf{F}_C = (-13.66 \text{ N}, -3.66 \text{ N}). \tag{180}$$

Alternatively, we could specify the magnitude and direction of $\mathbf{F}_C$. The magnitude is

$$F_C = \sqrt{F_{C,x}^2 + F_{C,y}^2} \tag{181}$$

$$= \sqrt{(-13.66 \text{ N})^2 + (-3.66 \text{ N})^2} \tag{182}$$

$$= 14.1 \text{ N}. \tag{183}$$

The direction of the force exerted by Cecil is found as $\theta = \arctan\frac{3.66}{13.66} = 15°$ away from the negative x-axis. See Figure 183.

FIGURE 183.
Solving Worked
Example 3. The
magnitude and
direction of the force
$\mathbf{F_C}$ that Cecil must
exert to cancel the
combined force of
Alice and Bob.

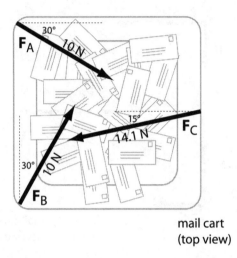

mail cart
(top view)

Another way to approach the problem would be to start by finding the net result of Alice and Bob's forces, then simply require that Cecil's push be equal and opposite to that.

So let's first add Alice's force $\mathbf{F_A}$ to Bob's force $\mathbf{F_B}$. We have already broken everything into components in Figure 182. Using these components, the combined force $\mathbf{F_{AB}}$ of Alice and Bob together has components

$$(184) \qquad F_{AB,x} = F_{A,x} + F_{B,x}$$

$$(185) \qquad = (+8.66\,\text{N}) + (+5\,\text{N})$$

$$(186) \qquad = +13.66\,\text{N}$$

and

$$(187) \qquad F_{AB,y} = F_{A,y} + F_{B,y}$$

$$(188) \qquad = (-5\,\text{N}) + (+8.66\,\text{N})$$

$$(189) \qquad = +3.66\,\text{N}.$$

This is a force with magnitude $\sqrt{(+13.66\,\text{N})^2 + (+3.66\,\text{N})^2} = 14.1\,\text{N}$ and direction $\arctan\frac{3.66}{13.66} = 15°$ above the positive x-axis

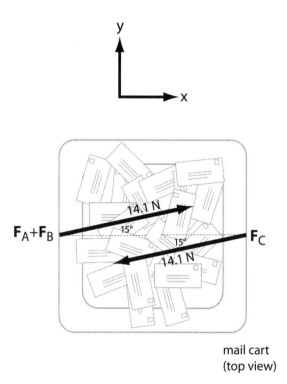

FIGURE 184.
Another way to solve
Worked Example 3.
Cecil must apply
a force equal and
opposite to $F_A + F_B$
in order to balance
the forces exerted by
Alice and Bob.

(Figure 184). A force from Cecil that would cancel $\mathbf{F}_A$ and $\mathbf{F}_B$ exactly would be an equal and opposite force: a force of magnitude 14.1 N directed 15° below the negative x-axis—just as we found using the first approach. ◇

The Effects of Applying Force at Different Places

If you've ever tried to move a piece of furniture, you know how much it matters where you grab hold! The strength of your push and the direction of your push are simply not the entire story. For example, consider two movers pushing a refrigerator with equal and opposite forces (Figure 185).

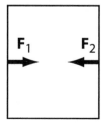

FIGURE 185.
Two movers
applying equal
and opposite
forces to a
refrigerator.

FIGURE 186.
Two movers exerting
the same forces as in
Figure 185, but
applying them at
different points on
the refrigerator.

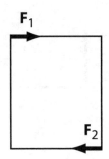

FIGURE 186.
Two movers exerting the same forces as in Figure 185, but applying them at different points on the refrigerator.

If the movers are applying their forces at the points shown, they will not accomplish much! But what if the movers apply the exact same forces at different points, as shown in Figure 186? You can probably see that the refrigerator in Figure 186 will have a tendency to *rotate*.

The refrigerator example illustrates two general principles. First, the point of application of a force does indeed determine whether the object will have any tendency to rotate. On the other hand, however, the point of application of a force does *not* matter if all you are interested in is the *translational* motion of the object—that is, the overall motion of the object as a whole from one place to another. Observe that even if the refrigerator in Figure 186 does begin to rotate, the center point of the refrigerator will stay put while the rotation occurs—just as it would if the refrigerator were subjected to the forces shown in Figure 185. With regard to the motion of the center point, the same forces always lead to the same result, no matter where on the object the forces are applied.

In this book, we shall not be analyzing the rotational dynamics of rigid bodies. We shall be concerned only with the translational motion of bodies as a whole. This means that for us it doesn't matter at all where the forces are applied to an object. Indeed, from now on, the force diagrams in this book will often show the target object as a single dot, with all of the applied forces shown as vectors with their tails on the dot. Issues of twisting, torsion, or rotation will then be invisible.

Reader's Notes

Below is space for you to summarize key points from this chapter.

FOCUSED PROBLEMS :: CHAPTER 11

At this point, it would be a good idea to go back and redo the Focused Problems from Chapter 4.

1. Two postal workers, Malik and Job, are pushing a massive mail cart across the floor. Malik is pushing with a horizontal force of 400 N toward the west wall of the post office. Job is pushing with a horizontal force of 300 N toward the north wall of the post office. What is the net force applied to the mail cart by Malik and Job?

2. Alice, Bob, and Cecil are trying to push a mail cart into DeeDee's office, but DeeDee doesn't want the mail cart because she'll have to sort all of that mail. So DeeDee is pushing back. The forces exerted by Alice, Bob, and Cecil are shown in Figure 187.

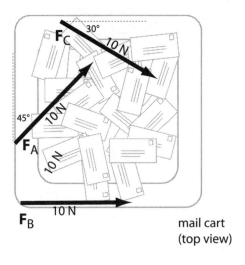

FIGURE 187.
Focused Problem 2.
The forces DeeDee
must balance to keep
the mail cart out of
her office.

mail cart
(top view)

What force $\mathbf{F}_D$ must DeeDee apply to the cart in order to cancel out the opposing forces exactly and create a situation with zero net force on the cart?

SOLUTIONS CAN BE FOUND ON PAGE 389

CHAPTER 12

"Newton's Little Law"

Congratulations! You've reached the very threshold of the System of the World. Already you have journeyed across a varied landscape of physics. You have studied rates of change, and also rates of change of rates of change. You have added and subtracted vectors. You have examined the meanings of position, velocity, and acceleration. You have adopted the physicist's definition of force as an influence of one object on another. And you have seen how forces applied to the same target combine to result in a single net force.

It is finally time to connect everything you know about motion to everything you know about force. The connection we are about to make is the connection Sir Isaac Newton first made in 1665. It is a connection that changed the world forever.

A Brief Refresher on Acceleration

Now would be a good time to reread Chapter 6. Go ahead—it's short. I'll wait until you get back.

What Forces Can Do

I don't watch too much football on TV, but I do like to daydream sometimes about what it would be like if you could broadcast a football game on TV using computers to make all of the players from one team invisible. It would look funny, wouldn't it? Imagine the quarterback looking around, not wanting to throw a pass to receivers who look totally open to us. Or linebackers waving their hands in a frenzy as though trying to swat invisible mosquitoes. Then imagine some powerful yet invisible influence totally smashing the quarterback to the ground. To the viewer at home, it would seem as though the players were up against a team of invisible and violent ghosts!

I like to imagine a scene in which a running back is barreling down the sideline at top speed, trying to score a touchdown. Then, *bam!* All of a sudden, for no apparent reason, he goes flying sideways into the bench! What's happened?

On the next play, the running back gets free again. He's running down the sideline and looks to be home free, when *bam!* He's suddenly stopped in his tracks.

On the next play, believe it or not, the running back gets free again. He's running down the sideline, and again he looks to be home free, when *bam!* This time he's propelled forward with unnatural speed, and he hurtles ahead through the air.

Obviously, what's happened in all of these cases is that the running back has been tackled. During each tackle, an opposing player has exerted a large force on the running back. Even though we haven't been able to see the opposing player, we can still get a pretty good idea of what happened in each case just by looking at the replays.

Replay 1. On this play, the running back got smashed sideways. His trajectory is shown in Figure 188. Obviously, what must have happened is that a huge force hit the running back from his left side and made him veer suddenly to his right.

Replay 2. On this play, the running back got stopped in his tracks. Figure 189 shows what must have happened: a huge backward force hit the running back and made him suddenly slow down.

Replay 3. On this play, the running back suddenly sped up and flew out of control. Figure 190 shows what must have happened: a huge forward force hit the running back and made him suddenly speed up.

There is a simple conclusion we can draw from these three scenarios:

(190)   External forces change your motion.

In each scenario, at first the running back was moving smoothly down the field, all on his own, headed for a touchdown. Then

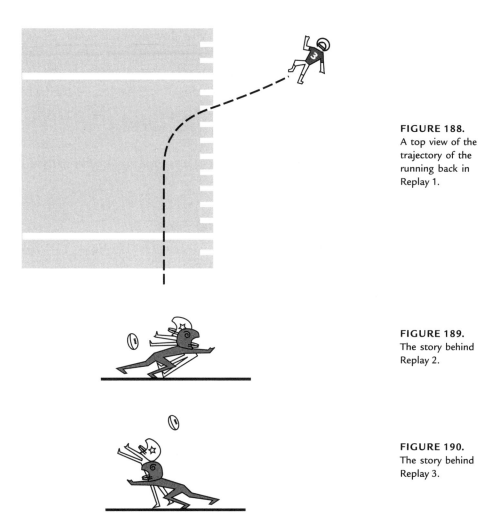

FIGURE 188.
A top view of the trajectory of the running back in Replay 1.

FIGURE 189.
The story behind Replay 2.

FIGURE 190.
The story behind Replay 3.

things suddenly *changed*. In each case, the motion of the running back changed in some way because a force was exerted on him by another player.

We can learn even more from this example: namely, *how* a force can change your motion. As the three scenarios illustrate, *a force can do one of three things:*

- speed you up (Replay 3),
- slow you down (Replay 2), or
- make you turn (Replay 1).

If you're standing still and someone pushes you, you speed up. If you're moving forward and someone pushes you from behind, you speed up. If you're moving forward and someone pushes you from the front, you slow down. If you're moving and someone pushes you from the side, you turn.

A force can speed you up, slow you down, or make you turn. There is a very concise way to put this. Have you guessed what it is?

(191)   Forces accelerate things.

That's because acceleration *means* speeding up, slowing down, or turning.

The Importance of the Net Force

Forces accelerate things . . . except when they don't. In Worked Example 3 of Chapter 11, Alice, Bob, and Cecil were all exerting forces on a mail cart. Although there were lots of forces acting on the mail cart, the mail cart was not accelerating. So were we wrong to say that forces accelerate things? No. All this example shows is that it's the *net* force that matters when it comes to making an object accelerate. The mail cart remained still because the forces exerted by each person were such that the *net* force on the cart canceled out to zero.

If you want to know how an object will accelerate, you have to know *all* of the significant forces that are acting on it. It is the confluence of all the external forces that determines the acceleration of the object.

"Newton's Little Law"

A fun way to summarize what we've decided so far is to imagine that, prior to publishing the *Principia* with its three Laws of Motion, Newton first published a preliminary draft of his thinking on things. In this first draft, we imagine Newton writing down the following statement:

(192) $\mathbf{F}_{net}$ and $\mathbf{a}$ point the same way.

Newton's Little Law: $\mathbf{F}_{net}$ and $\mathbf{a}$ point the same way.

This statement has a nice sing-song rhythm that makes it easy to remember. Say it with me: "$\mathbf{F}_{net}$ and $\mathbf{a}$ point the same way."

I call Equation 192 Newton's Little Law. It's not the full Second Law of Motion, which we shall study in the next chapter, but it's incredibly useful all on its own. Spelled out in full, Newton's Little Law says:

The net force vector acting on an object
at any given moment points the same
way as the object's acceleration vector
at that same moment.

(193)

This is a wonderful revelation! Think back to Chapter 6, where we sketched acceleration vectors for moving objects. Thanks to Newton's Little Law, every time we sketched the acceleration vector of a moving object, we were also basically sketching the net force acting on the object at that moment! Newton's Little Law allows us to use the observed motion of an object to infer a great deal about the forces that are at work on it. This is our first glimpse of the way Newton's Laws allow us to uncover the hidden structure of the universe (Chapter 14).

To show how this works, we continue Worked Example 4 from Chapter 6.

»WORKED EXAMPLE 1

A clown is shot out of a cannon. Earlier we sketched velocity and acceleration vectors for the clown based on reasoning about speeding up, slowing down, and turning. The velocity and acceleration vectors are shown in Figure 191. At each of the moments (i), (ii), and (iii), sketch the net force acting on the clown.

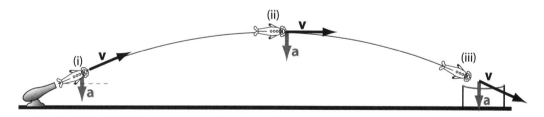

FIGURE 191. Worked Example 1. Velocity and acceleration vectors at different times for a clown shot from a cannon, based on reasoning about speeding up, slowing down, and turning.

»SOLUTION 1

By Newton's Little Law, $\mathbf{F}_{net}$ and $\mathbf{a}$ point the same way. The $\mathbf{F}_{net}$ vectors are shown in Figure 192.

By carefully analyzing the motion of the clown in terms of speeding up, slowing down, and turning, we have managed to divine the direction of the net force acting on the clown at each

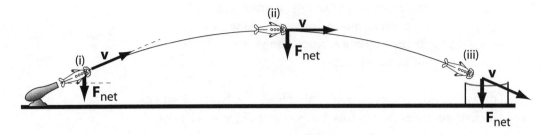

FIGURE 192.
Solution to Worked
Example 1. The net
force acting on the
clown in Figure 191
at each of the indi-
cated moments of
time.

moment of time. In this case, it appears that the net force acting on the clown is always more or less straight down. Any guesses as to what object is exerting this force on the clown? ◇

Here's another example of how we can apply Newton's Little Law.

»WORKED EXAMPLE 2

You are pushing a grocery cart down the supermarket aisle. During the entire trip, the cart moves in a straight line at a constant speed. What is the net force acting on the cart during this time?

»SOLUTION 2

The cart is not speeding up, slowing down, or turning. Therefore, the cart's acceleration vector is zero. Therefore, by Newton's Little Law the net force acting on the cart must also be zero the whole time! ◇

Many of my students think this is a crazy answer. First, they say, "I *know* that I'm pushing on the cart, so how can the net force on the cart be zero?" And second, "How can something be moving with no net force acting on it?"

Let's take the first point first. You are indeed pushing on the cart. But you're not the *only* thing pushing on the cart, are you? Notice that the cart is also in contact with the floor. Apparently the floor is pushing backward on the cart with a force equal and opposite to the force of your own push. See Figure 193.

You may not be thinking about what the floor is doing to the cart as you walk down the aisle, but you can easily demonstrate the floor's influence: Simply let go of the handle, and the floor will bring the cart to a stop.

Figure 193 answers the first question students have, "How can the net force on the cart be zero even though I am pushing on

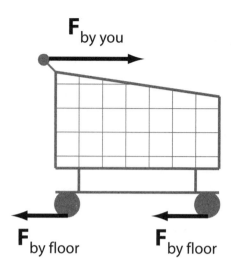

F_{by you}

F_{by floor}   **F**_{by floor}

FIGURE 193.
Solution to Worked Example 2. Horizontal forces on a shopping cart being pushed down the grocery aisle in a straight line at constant speed.

it?" (Answer: The floor is pushing on the cart, too, so the net force cancels out to zero.)

Now for the second question: "How can something be moving when there is no net force acting on it?"

I think students ask this question for many different reasons. One of the most common reasons I see actually has nothing to do with force at all! It's really a problem with describing motion. This problem arises when students fail to distinguish between *getting* in motion and *remaining* in motion.

Unfortunately, physics problems are usually set up to wreak havoc with this confusion because the "scenes" presented in physics problems so often begin *after* the "getting in motion" has already happened. The shopping cart problem is an example of this. To get your cart to the snack aisle in the first place, you must earlier have taken the shopping cart from a rack in front of the store, where it was initially at rest. Therefore, at some point you obviously did have to *get* the shopping cart in motion. However, the problem stated in Worked Example 2 ignores all of these prior events. It's like what happens when a film director cuts to a scene in which the action is already in progress. When the scene in Worked Example 2 begins, the cart is already in motion. The speeding up of the cart happened before the scene began. And the net force responsible for that speeding up disappeared as soon as the speeding up was complete.

Are you startled to read that forces can disappear? Perhaps you should reread Chapter 10.

Influence versus Inertia: External versus Internal

Once we've cleared away the simple misconceptions about the shopping cart problem—the fallacy of thinking that *your* push on the cart is the *only* push and the confusion between *getting* in motion and *staying* in motion—the deeper issues can emerge. Now we really know what we're asking when we ask, "How can the shopping cart keep moving when there is no net force acting on it?"

One reason students continue to ask this question, even after the simple misconceptions have been cleared away, is that they have not yet really absorbed the idea that *a force is an influence of one object on another.* If you see a shopping cart trundling down the grocery aisle, minding its own business, moving in a straight line at a constant speed, can you really believe that the cart is under the influence of some *other* object? What object would that be, pray tell? In fact, I would say that an object moving along in a straight line at a constant speed is the very image of something that is *not* being influenced by anything else! Think back to the example of the running back. You didn't infer any outside influence when he was running in a straight line at a constant speed. You inferred the outside influence when he was suddenly stopped in his tracks!

Or again, picture a tiny pebble out in deep space, drifting along in a straight line at a constant speed. The acceleration of the pebble is zero. So Newton's Little Law implies that the net force acting on the pebble must also be zero. But even without applying Newton's Little Law, you should recognize that it would be absurd to try to explain the persistent motion of this solitary pebble by invoking some kind of external influence. This pebble is the very image of *non-influence.* There is nothing around to influence it! For this reason, force per se cannot be what is keeping the pebble moving.

But the question remains: What *does* keep the pebble moving?

If you accept that no *other* object is keeping the pebble moving—that this is not a case of outside influence, or force, in any sense—you must conclude that *something in the pebble itself,* something besides force, keeps the pebble moving. This innate *something*—this urge, which every object has, to keep moving in a straight line at a constant speed, absent any external influences—is what Newton called "inertia." Newton didn't explain *why* objects in our universe have inertia, but he knew enough to distinguish between an object's *innate* desire to keep moving and the *external* influences that tend to *change* an object's motion.

Newton's First Law

Seeing as we're talking about inertia now, I think this would be a good time to state Newton's First Law, sometimes called "the Law of Inertia." Newton's First Law is a very tiny sliver of the Laws of Motion. It deals only with situations like those of the pebble in outer space and the unmolested running back. Here it is:

> *Newton's First Law*: If all of the forces acting on an object balance out to zero, the object will either move in a straight line at a constant speed or remain at rest over time. Conversely, if an object is moving in a straight line at a constant speed or isn't moving at all, the forces on the object must be balancing out to zero.

The First Law follows mathematically from Newton's Little Law. For because $\mathbf{F}_{net}$ and $\mathbf{a}$ point the same way, $\mathbf{F}_{net}$ equals zero when and only when $\mathbf{a}$ equals zero—that is, when the velocity vector remains constant—that is, when the object moves in a straight line at a constant speed or remains at rest [5].

The Weaker Link between Force and Velocity

According to Newton's Little Law, "$\mathbf{F}_{net}$ and $\mathbf{a}$ point the same way." The connection between force and acceleration couldn't be simpler. What about the link between force and velocity? Is there a link at all?

The answer is that *over time* there's a link. If you apply a steady force $\mathbf{F}_{net}$ to an object for a long enough time, its velocity vector $\mathbf{v}$ will *eventually* turn itself around more and more to point along $\mathbf{F}_{net}$. You can see this happening in Figure 192. The net force on the clown is always pointing downward. *Over time,* this downward force acts to rotate the velocity vector from its initial upward direction to a direction more and more downward-pointing. However, at any *fixed* instant, there is no obvious relationship between the $\mathbf{F}_{net}$ and $\mathbf{v}$ vectors.

Students often want their force diagrams to show them something about how their target is moving at the moment of time in question. But force diagrams can't do that. Indeed, because $\mathbf{F}_{net}$ points along the acceleration vector rather than the velocity vector, it would be better to say that the force diagram shows you something about how the target is *about to move*. This observation gives us just a glimpse of the way Newton's System of the World allows us to predict the future from the past. (More on this in Chapter 14.)

Reader's Notes

Below is space for you to summarize key points from this chapter.

FOCUSED PROBLEMS :: CHAPTER 12

1. Figure 194 shows three force diagrams for three different targets, (a), (b), and (c). No significant forces are missing from any of the diagrams. In each case, the target's velocity vector is also shown at the moment of time being analyzed. Which object(s) are speeding up at this moment? Which are slowing down at this moment? Which are turning at this moment?

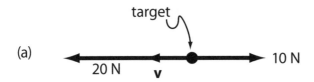

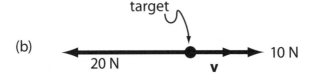

FIGURE 194.
Focused Problem 1.
Force diagrams for
the targets (a), (b),
and (c).

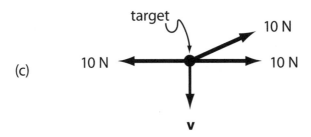

2. Draw force diagrams for the following situations, familiar from Chapter 6:
   a. A tiny tot swinging in a swing, at the moment when the tot is at the bottom of her trajectory.
   b. Same tot a few moments after reaching the bottom of her trajectory.

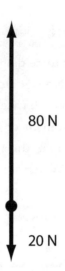

FIGURE 195.
Focused Problem 3.
All of the significant
forces acting on a
bird at a certain
instant of time.

80 N

20 N

c. Same tot a few moments before reaching the bottom of her trajectory.

d. An elevator at a moment when it is moving upward and speeding up.

e. Same elevator at a moment when it is moving upward at a constant speed.

f. Same elevator at a moment when it is slowing to a stop (still moving upward).

g. Same elevator at a moment when it is moving downward and speeding up.

h. Same elevator at a moment when it is moving downward at a constant speed.

i. Same elevator at a moment when it is slowing to a stop (still moving downward).

In each case, verify for yourself that the net force on the target points in the same direction as the target's acceleration vector.

Also, for each case, answer this question: Does or doesn't the net force point in the same direction as the target's direction of motion?

3. Figure 195 shows all of the significant forces acting on a bird at a certain instant of time.

a. Sketch a vector representing the net force acting on the bird.

   b. Which way is the bird moving at this instant of time? (That is, in what direction does the bird's velocity vector point at this instant of time?) Up, down, left, or right?

   c. Could the bird be at rest at this instant of time? Explain.

4. Think of real-life situations that match each of the following scenarios:

   a. The net force on an object points to the left, and the object is moving to the right.

   b. The net force on an object points down, and the object is moving to the right.

   c. There is no net force on an object, and the object is moving straight up.

   d. The net force on an object points up, and the object is moving straight up.

   e. The net force on an object points up, and the object is at rest.

    For each scenario, draw
- a cartoon of the situation
- a force diagram for the target
- a separate sketch of the net force vector
- a separate picture showing the target's velocity and acceleration vectors.

    For each of the situations you thought of, say whether the object is moving on a curved path or not at the instant of time that you are considering it. Also, is the object speeding up? Slowing down?

    For each of the situations you thought of, does your net force vector point in the same direction as your acceleration vector? (It should!)

5. Here's a theory about the earth's motion around the sun: The earth moves around the sun *at a constant speed,* in an elliptical orbit, with the sun at the center of the ellipse. To see what we make of this theory, draw the earth's orbit, showing the earth at six different positions throughout the year. Sketch the net force acting on the earth at these six times. Assume that the

sun is the only object exerting a force on the earth. What don't you like about the force vector in my theory?

6. Figure 196 shows a hawk gliding in for a landing. Let us suppose that she glides down in a straight line and that her speed is constant at 5 m/s as she descends.

FIGURE 196.
Focused Problem 6.
A hawk gliding in for
a landing.

5 m/s

Note that only two forces act on the hawk during her descent:

- a gravitational force from the earth, $F_{by\ earth}$
- a force from the air, $F_{by\ air}$.

a. Draw a force diagram for the hawk during her descent.

Make your force diagram to scale in the sense that if one of these two forces is larger than the other, the arrow from the larger force is noticeably larger, while if the two forces are equal, the arrows from the forces are about equally long. Also, be careful to draw the force vectors so that they point in the direction in which you mean them to point.

b. Fully explain how you decided which force, if either, was larger in magnitude than the other.

c. Fully explain how you decided the direction in which $F_{by\ air}$ points.

7. Figure 197 shows a helicopter flying upward at an angle. The helicopter flies in a straight line, and its speed is constant at 5 m/s as it ascends.

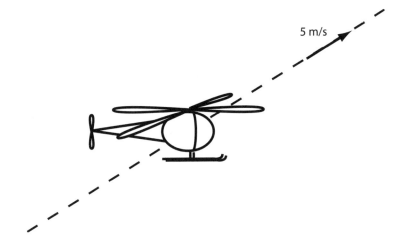

5 m/s

FIGURE 197.
Focused Problem 7. A
helicopter ascending.

Note that only two forces act on the helicopter during its ascent:

- a gravitational force from the earth, $F_{\text{by earth}}$
- a force from the air, $F_{\text{by air}}$.

a. Draw a force diagram for the helicopter during its ascent.

Make your force diagram to scale in the sense that if one of these two forces is larger than the other, the arrow from the larger force is noticeably larger, while if the two forces are equal, the arrows from the forces are about equally long. Also, be careful to draw the force vectors so that they point in the direction in which you mean them to point.

b. Fully explain how you decided which force, if either, was larger in magnitude than the other.

c. Fully explain how you decided the direction in which $F_{\text{by air}}$ points.

8. A shopper at a mall is riding up an escalator. Sketch a force diagram for the shopper. Sketch the shopper's velocity and acceleration vectors. Is everything in accordance with Newton's Little Law? Try this problem two ways: First, neglect the small force of air resistance on the shopper. Second, include this force in the analysis.

SOLUTIONS CAN BE FOUND ON PAGE 389

CHAPTER 13

Newton's Second Law

In the previous chapter we examined "Newton's Little Law," which says that $\mathbf{F}_{net}$ and $\mathbf{a}$ point the same way. In this chapter we'll refine Newton's Little Law to take into account the masses of target objects.

Mass

"What is mass?" is a profound question. Indeed, modern physical theories such as field theory and string theory are still trying to sort this out. In this book we'll take the commonsense view Newton himself took: Mass is simply the amount of "material stuff" contained in an object. Mass is measured in kilograms (kg). In everyday terms, a quart of milk contains about a kilogram of mass. A large person contains about a hundred kilograms of mass.

Today the official basis for measuring mass is the *standard kilogram,* a machined cylinder of platinum-iridium alloy housed in Sèvres, France, at the Bureau International des Poids et Mesures (International Bureau of Weights and Measures). However, alternatives to this approach are currently being discussed, such as defining a kilogram as the amount of mass contained in a certain agreed-upon number of carbon atoms.

The Effect of Mass on Acceleration

A force of magnitude 100 N is strong enough to lift a bag of groceries. But it's not strong enough to lift a car. Obviously, the capacity of a force to alter the motion of an object has something to do with how massive the object in question is.

Think about a race between two parents pushing baby carriages. The parents are equally matched in terms of their strength. But one parent has quintuplets sitting in a five-seater carriage, while the other parent has just one baby sitting in a single-seater carriage. Which team do you think is going to have the better acceleration off the starting blocks?

The single-baby team will obviously accelerate at a higher rate, given the same strength of push from the parent. I like to think of this in terms of input and output. If we apply an "input" force of a certain strength, we achieve an "output" acceleration of a certain magnitude. Increasing the applied force will increase the resulting acceleration. So there seems to be a proportionality here:

$$\mathbf{a}_{output} = (\text{proportionality constant})\mathbf{F}_{net\ input}. \tag{194}$$

The proportionality constant depends on what the target object is. If we are talking about a bag of groceries, an input net force of 100 N will generate a noticeable output acceleration, whereas if we are talking about a boulder, an input net force of 100 N will generate only a small output acceleration.

Based on this single observation, we might well guess that the output acceleration resulting from a given input force scales *inversely* with the mass of the target. And that is indeed the content of Newton's Second Law:

$$\mathbf{a} = \frac{1}{m}\mathbf{F}_{net}. \tag{195}$$

Or, in words: The acceleration of an object is one over the mass of the object times the net force acting on the object. The mass is in the denominator to ensure that a given applied force leads to an acceleration that is small when the mass of the target object is large.

Newton's Second Law: $\mathbf{a} = \frac{1}{m}\mathbf{F}_{net}$. At each instant of time, the acceleration vector of an object is equal to one over the mass of the object times the net force acting on the object at that time.

How Newton's Second Law Relates to "Newton's Little Law"

The inverse mass $\frac{1}{m}$ is always a positive number—a positive *scalar*, in the language of Chapter 4. So when we multiply the net force vector $\mathbf{F}_{net}$ by this scalar according to Equation 195, the mathematical effect is to leave the direction of the vector unchanged. We see once again that $\mathbf{F}_{net}$ and $\mathbf{a}$ point the same way. Newton's Second Law implies "Newton's Little Law."

The Units of Force

Although multiplying the vector $\mathbf{F}_{net}$ by $\frac{1}{m}$ preserves the direction of $\mathbf{F}_{net}$, the new vector does have a new magnitude with new units. The units of $\frac{1}{m}\mathbf{F}_{net}$ will be $\frac{1}{kg} \times N$. But because $\frac{1}{m}\mathbf{F}_{net} = \mathbf{a}$, with units of m/s^2, we will have $\frac{1}{kg} \times N = m/s^2$, or

$$N = \frac{kg\,m}{s^2}. \tag{196}$$

We see that 1 newton is just shorthand for $1\frac{kg\,m}{s^2}$.

Rearranging Newton's Second Law

Let's multiply both sides of Equation 195 by m and interchange the right-hand side with the left-hand side. This gives

$$\mathbf{F}_{net} = m\mathbf{a}. \tag{197}$$

Equation 197 is the traditional formulation of Newton's Second Law. This version is useful if you want to do what Newton himself did in several cases: observe the acceleration of a moving object such as the moon, and thereby infer the forces that must be acting on it. In this endeavor, it is the $\mathbf{a}$ that is the "input," whereas the $\mathbf{F}_{net}$ is the "output." We'll have more to say on the different uses of $\mathbf{F}_{net} = m\mathbf{a}$ and $\mathbf{a} = \frac{1}{m}\mathbf{F}_{net}$ in Chapter 14.

"Newton's Little Law" from Chapter 12 was a *non-numerical* statement of how the net force vector relates to the acceleration vector: namely, they point the same way. Newton's Second Law in its full form now allows us to make *numerical* connections between these vectors. For example, because the vector $\mathbf{F}_{net}$ is equal to the vector $m\mathbf{a}$, it follows that the magnitudes of these vectors are also equal. In other words,

$$F_{net} = ma \quad \text{and} \quad a = \frac{1}{m}F_{net}. \tag{198}$$

> Newton's Second Law: $\mathbf{F}_{net} = m\mathbf{a}$. At each instant of time, the net force acting on a body is equal to the mass of the body times the acceleration vector of the body.

These are true statements about the strength of the net force and the magnitude of the acceleration, regardless of the common direction of the two vectors.

» WORKED EXAMPLE 1

The moon has mass $M = 7.4 \times 10^{22}$ kg and orbits the earth once every 28 days with a more or less constant speed in a more or less circular path of radius $R = 3.8 \times 10^{8}$ m. Find the net force acting on the moon at any given time.

» SOLUTION 1

We know enough about the moon's motion to calculate the magnitude of its acceleration vector. Then, using Newton's Second Law, we can infer the value of the net force acting on the moon!

The moon's velocity and acceleration vectors are shown at four instants of time in Figure 198, along with the net force vectors at these times (which I simply drew so that they point the same way as **a**). Distances in this diagram are not to scale.

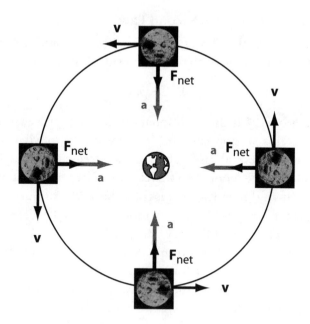

FIGURE 198. Worked Example 1. The moon's velocity and acceleration vectors at four different times, along with the net force vectors acting on the moon at those times. (The famous image of the moon is from the 1902 film *Le voyage dans la lune,* by Georges Mélières.)

The moon is not speeding up and not slowing down, so this is a simple case of pure turning. The acceleration vector therefore has magnitude $a = a_\perp = \frac{v^2}{r}$. Because the speed v and the radius r are constant in time, the magnitude of the acceleration is constant in time. And because the magnitude of the acceleration vector is constant in time, the magnitude of the net force vector is constant in time.

To find the value of $a_\perp$ we need to know the speed of the moon. Using simple distance-over-time reasoning, we note that the moon travels a circular distance $2\pi R = 2.39 \times 10^9$ m over a period of 28 d $= 2.42 \times 10^6$ s, giving a speed of 990 m/s. The perpendicular (and only) component of the acceleration vector is therefore $a_\perp = (990 \text{ m/s})^2/(3.8 \times 10^8 \text{ m}) = 0.0026 \text{ m/s}^2$. So, according to Newton's Second Law, the net force acting on the moon is of magnitude $F_{\text{net}} = (7.4 \times 10^{22} \text{ kg})(0.0026 \text{ m/s}^2) = 1.9 \times 10^{20}$ N.

A force of 100 billion billion newtons bends the moon's trajectory into a circular path! ◇

» WORKED EXAMPLE 2

At time $t = 0$, you find yourself back in the grocery store, standing at one end of the snack aisle with your cart at rest in front of you. After gathering your courage for a few seconds, you begin pushing the cart at time t_1 with a constant force of 100 N; this is enough of a push to overcome the 10 N frictional force exerted on the cart by the floor. The cart speeds up from time t_1 to time t_2, at which point the cart attains its maximum speed. Now you run along behind the cart for a few moments, lightening your push so that you merely cancel out the 10 N frictional force. The cart therefore proceeds with a constant speed from time t_2 to t_3, at which point you let go of the cart and stop running. The cart, now subjected only to the 10 N frictional force, slows down, coming to rest at time t_4. You stand there watching the stationary cart from time t_4 to t_5.

(i) Sketch a force diagram showing the relevant forces acting on the cart at a specific moment of time somewhere between time $t = 0$ and time $t = t_1$. Then make four more force diagrams illustrating specific moments between t_1 and t_2, between t_2 and t_3, between t_3 and t_4, and between t_4 and t_5. (ii) On each force diagram, draw the velocity and acceleration vectors of the cart. (I'd use a different color ink so that these vectors don't look like forces themselves!) In each diagram, does the net force vector point the same way as the acceleration vector? Does the relationship between **a** and **v** faithfully record the facts as to whether the cart is speeding up, slowing down, or turning at the moment shown? (iii) With the x-axis drawn so that it runs along the snack aisle in the direction of the cart's motion, sketch rough non-numerical $a_x - t$, $v_x - t$, and $x - t$ graphs for the cart over the range $t = 0$ to $t = t_5$. (iv) Sketch a rough non-numerical graph of $F_{\text{net},x}$ over time. (v) If the cart has a mass of 7 kg, label as many of your graphs as possible with numerical values.

» SOLUTION 2

If the net force vector ever had any vertical component, the acceleration vector would also have to have a vertical component. This would translate into a curvature of the cart's trajectory. In simple terms, we would see the cart lift off the ground or sink into it!

(i) My force diagrams are shown in Figure 199. I have not bothered to show any vertical forces, because if the supermarket floor is level, the vertical forces acting on the cart must all cancel out anyway.

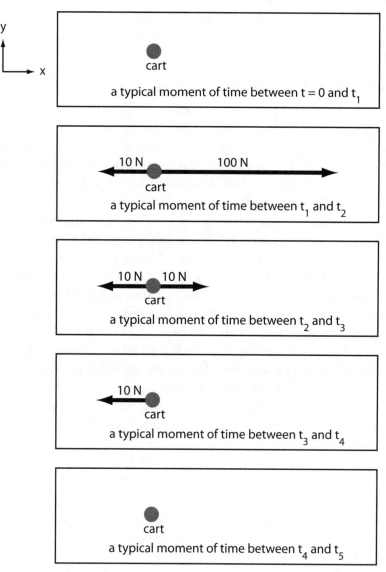

FIGURE 199. Solution to Worked Example 2 (i). Force diagrams for the shopping cart at selected moments of time.

(ii) In Figure 200, I have added the cart's velocity and acceleration vectors at each of the illustrated moments of time.

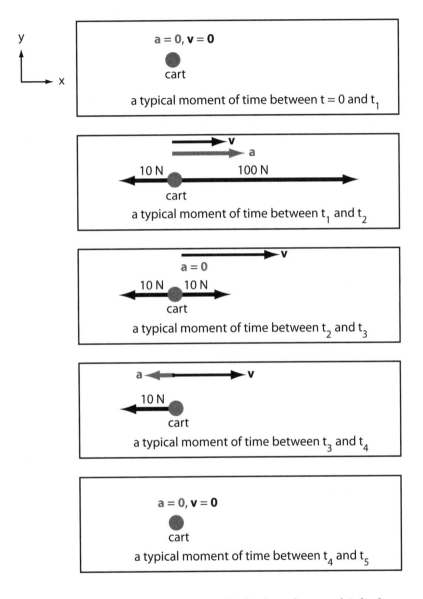

FIGURE 200. Solution to Worked Example 2 (ii). The force diagrams from Figure 199 augmented with the cart's velocity and acceleration vectors.

To draw the velocity vectors, all I had to do was think about which way the cart was moving, and how fast. I did this just by rereading the given scenario.

To draw the acceleration vectors, all I had to do was think about whether the cart was speeding up, slowing down, or turning at each moment. Again, I did this just by rereading the given scenario. If the cart was speeding up, I drew my **a** vector along the **v** vector.

If the cart was slowing down, I drew my **a** vector *against* the **v** vector. If the cart was neither speeding up nor slowing down—and of course it wasn't turning, either—I wrote **a** = **0**.

In drawing my acceleration vectors, I did not pay particular attention to the force vectors on my diagrams. But now I had better make sure that the directions of my acceleration vectors agree with the directions of my net force vectors. A glance at each force diagram shows that everything works out: when the acceleration vector is pointing right, so is the net force vector. When the acceleration vector is pointing left, so is the net force vector. And when the acceleration vector is zero, so is the net force vector.

Sometimes in solving problems you will draw your acceleration vectors based on a knowledge of how the object is moving. Thanks to Newton's Second Law, this will tell you a lot about the net force acting on the object. Conversely, sometimes you will draw your acceleration vectors based on a knowledge of the forces acting on the object; this will tell you a lot about how the object is moving. In the most difficult sorts of problems, you won't know everything about either the motion of the object *or* the forces that act on it, and you'll have to use the Second Law to analyze all aspects of the problem simultaneously.

(iii) My graphs are shown in Figure 201. I drew the $a_x - t$ graph first. I made a_x positive or negative based on which way the acceleration vectors were pointing in Figure 200—with the arrowhead of the x-axis or against it.

I made the acceleration constant during each interval because the forces acting on the cart were themselves unchanging during each interval. And if $F_{net,x}$ remains constant, a_x must also remain constant.

With the $a_x - t$ graph in hand, I knew enough to draw the $v_x - t$ graph. This I did by making sure that my $v_x - t$ curve always had a slope (rate of change) equal to the value shown in the $a_x - t$ graph. I also allowed myself to be guided by the description of the scenario, which indicated the times when the cart was at rest and the times when it was moving fast.

Is it time for another visit to Chapters 2 and 9?

With the $v_x - t$ graph in hand, I could draw the $x - t$ graph. This I did by making sure that my $x - t$ curve always had a slope (rate of change) equal to the value shown in the $v_x - t$ graph.

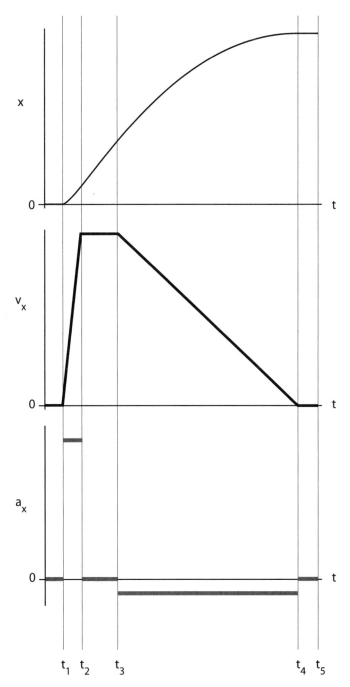

FIGURE 201. Solution to Worked Example 2 (iii). Illustrating the motion of the cart with $x - t$, $v_x - t$, and $a_x - t$ graphs.

(iv) A wonderful thing about Newton's Second Law is that once you've sketched a rough non-numerical $a_x - t$ graph, you've also sketched a rough non-numerical $F_{net,x} - t$ graph! Because $F_{net,x} = ma_x$, these two quantities are proportional, with a positive scale factor. This means that the overall shape of the graph doesn't change. My $F_{net,x} - t$ graph is shown in Figure 202.

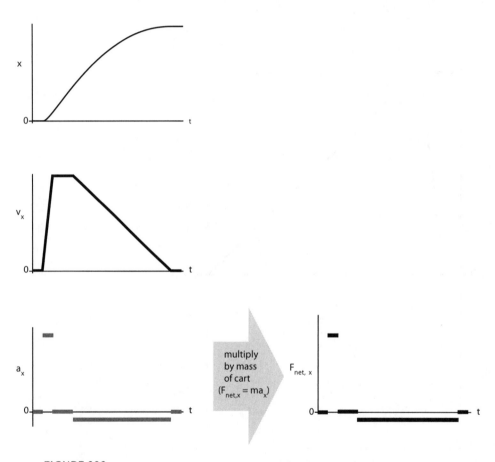

FIGURE 202.
Solution to Worked Example 2 (iv). Graphing the net x-force acting on the cart by multiplying the a_x graph by the mass of the cart.

(v) During the period t_1 to t_2, the net force acting on the cart has magnitude 90 N. See Figure 199. Knowing that the cart has a mass of 7 kg, we can compute the magnitude of the cart's acceleration during this phase: $a = F_{net}/m = 90\,\text{N}/7\,\text{kg} = 12.9\,\text{m/s}^2$. Likewise, during the period t_3 to t_4, the net force acting on the cart has magnitude 10 N. This gives an acceleration of magnitude $10\,\text{N}/7\,\text{kg} = 1.4\,\text{m/s}^2$. The two acceleration values we just calculated are shown in Figure 203.

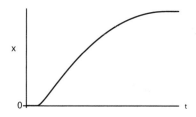

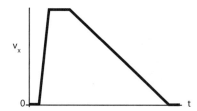

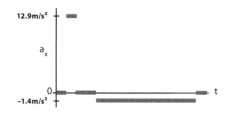

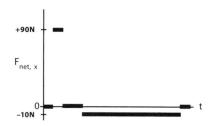

FIGURE 203.
Solution to Worked Example 2 (v). The graphs labeled with correct numerical values.

Finally we can return to the $F_{\text{net},x} - t$ graph and label the vertical axis with our two values, $+90$ N and -10 N. This is also shown in Figure 203. ◇

Accelerating Frames of Reference: A Caution

Newton's Second Law allows us to explain and predict an incredible range of phenomena. But there is one caveat. The Second Law explains observed phenomena correctly only when the observer is not accelerating. Let me give an example of how the whole enterprise breaks down when we try to use the Second Law within an accelerating frame of reference.

There is a U.S. government filmstrip that shows Air Force Colonel Dr. John Stapp riding a rocket sled. Three frames from the film are shown in Figure 204. The first frame shows Dr. Stapp at rest. The second frame shows Dr. Stapp moving forward and speeding up at a stupendous rate. The third frame shows Dr. Stapp

FIGURE 204.
Three frames from a film clip of Dr. John Stapp undergoing accelerations of great magnitude. (From www.af.mil/ news/airman/0498/ sled4.htm.)

In this frame, it looks as if something is pulling Dr. Stapp's skin backward.

No object is pulling it backward, though.

In this frame, it looks as if something is pulling Dr. Stapp's skin forward.

No object is pulling it forward, though.

moving forward and slowing down at a stupendous rate. The effects on his face are amazing!

Focus on the second frame. There's something strange about this image. The flesh on Dr. Stapp's face seems to be *pulled backward*.

Why do I call this strange? Well, because I can't identify any objects whatsoever that are actually *pulling* Dr. Stapp's face backward. What on earth could be doing this? See Figure 205 for one suggestion.

FIGURE 205.
A gremlin that might be pulling Dr. Stapp's face backward. But probably not.

Common sense tells us that there is no object pulling Dr. Stapp's skin backward. Common sense is right. There is no backward force on his skin. And yet this film shows us that his skin was being pulled back.

There appears to be a force involved . . . but there isn't one. This paradox cannot be explained using Newton's concept of force at all. The trouble is that the film of Dr. Stapp was made by an *accelerating camera*. The camera was accelerating because it was fixed to the rocket sled, which was itself accelerating. Images from an accelerating camera can't be explained using Newton's Second Law.

Newton's whole take on forces and motion breaks down in an accelerating frame of reference. Newton himself recognized this perfectly well. That's why in his *Principia* he prefaced the Laws of Motion with a discussion of frames of reference, including a discussion of how to tell when you're in an accelerating frame of reference.

So is the Second Law useless for understanding what's going on with Dr. Stapp's cheeks? Not at all. It's just that in order to apply Newton's thinking, you have to make a mental leap and view the events from the perspective of someone standing on the ground. So let's agree to view these events from the ground. If we were there, we would see Dr. Stapp's cheeks—along with the rest of him—accelerating forward. Applying Newton's Little Law—which it's safe to do now, because we are viewing events from the ground—we would conclude that the net force on his cheeks must also be forward. But what *object* is exerting this forward force on his cheeks? His cheekbones! See Figure 206. The force on his cheeks is forward, which matches the forward acceleration we observe while standing on the ground. Everything is in accordance with the Second Law.

To explain phenomena using Newton's force concept and his Second Law, "put yourself on the ground," where you can see the accelerations of objects in their true light.

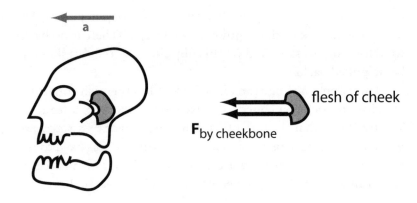

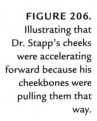

FIGURE 206.
Illustrating that Dr. Stapp's cheeks were accelerating forward because his cheekbones were pulling them that way.

Now for another example. When you're in the passenger seat of a car and the driver makes a hard left turn, you may sense yourself being "forced" to your right. But is any object actually pushing you to your right? Do you feel pressure on your left shoulder, for instance? No. Obviously, then, there is no object pushing you to your right. Indeed, where you feel pressure is on your *right* shoulder—evidence that it is the car door that creates your acceleration by pushing you *leftward*, "herding" you into a circular path the way the earth was herding the moon into a circular path in Worked Example 1 of this chapter. Or, in an analogy with the example of the running back, think of yourself as being "slowly tackled" by the car door. See Figure 207. Everything makes sense as long as we view events from a nonaccelerating perspective.

Here's another example. In the movie *Pirates of the Caribbean 2: Dead Man's Chest,* there is a wonderful scene in which the hero, Captain Jack Sparrow (played by Johnny Depp), is sitting inside a water wheel that has broken loose and is now rolling downhill. To shoot the scene, the director, Gore Verbinski, mounted a camera inside the wheel, trained on Depp's face. In the scene, Sparrow is wearing a key on a string around his neck, and there is a brief instant when we see the key magically begin to float upward off Sparrow's chest and levitate before his face! None of us thinks that keys can magically float. We automatically realize that in order to understand this observation, we have to mentally step back and watch the wheel from the ground. There we will see that everything can be explained in terms of gravitational forces and Newton's Second Law.

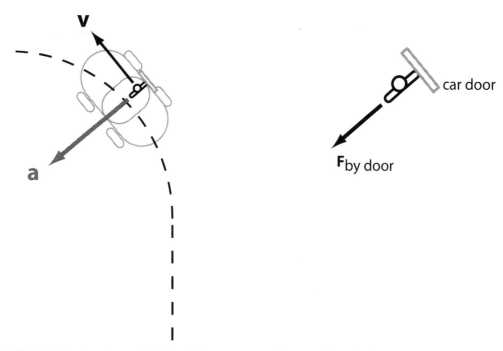

FIGURE 207. Being forced to your left when riding as a passenger in a car making a hard left turn.

Finally, let me point you to a funny illustration of accelerating reference frames on YouTube. See www.youtube.com/watch?v=3WyULc3jCgg.

Reader's Notes

Below is space for you to summarize key points from this chapter.

FOCUSED PROBLEMS :: CHAPTER 13

1. Two postal workers, Dallas and Todd, are pushing a massive mail cart across the floor. The cart rolls across the floor with negligible friction. Dallas is pushing with a horizontal force of 400 N toward the west wall of the post office. Todd is pushing with a horizontal force of 300 N toward the north wall of the post office.

   a. If the mail cart has a mass of 150 kg, what is the magnitude of the acceleration of the mail cart?

   b. What is the direction of the cart's acceleration vector?

   c. How fast is the cart moving? In what direction?

2. This problem refers to Figure 208.

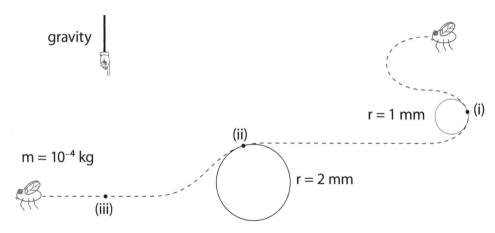

gravity

$m = 10^{-4}$ kg

r = 1 mm

r = 2 mm

(i)

(ii)

(iii)

FIGURE 208.
Focused Problem 2.
A fly in flight at three moments of time.

   a. For each of the three moments (i), (ii), and (iii) indicated in Figure 208, sketch the net force vector acting on the fly, assuming that the fly is speeding up at (i) and (ii) but neither speeding up nor slowing down at (iii). Also, for each of the three moments in time, draw a plausible force diagram for the fly. Assume that the earth always pulls straight down on the fly. What object(s) are responsible for the other force(s) in your diagram?

   b. In Table 4, fill in the numerical values that belong in the empty cells.

TABLE 4. Numerical Values for Focused Problem 2

Moment	v	$a_\parallel$	r	$a_\perp$	a	F_{net}
(i)	0.1 m/s	0.05 m/s²	1 mm			
(ii)		0.05 m/s²	2 mm		0.08 m/s²	
(iii)	1 m/s	0				

3. For Worked Example 2 of this chapter (see especially Figure 203), calculate the peak speed reached by the shopping cart. Also calculate the total distance traveled by the shopping cart during the entire episode. Use the numerical values given in the Worked Example. Also take the duration of the interval $t_1 \to t_2$ to be 1 s, and take the duration of the interval $t_2 \to t_3$ to be 2 s. (This is enough information for you to do all I've asked.)

4. In Worked Example 1 of Chapter 12, a clown is shot out of a cannon. Assume that during the time when the clown is in flight, the only force acting on him is a downward force of 980 N that remains constant in both magnitude and direction during his entire flight. Assume that the clown has a mass of 100 kg.

    The clown is shot toward the right. Lay down a coordinate system with the x-axis directed to the right and the y-axis directed upward. Let the origin be the tip of the cannon from which the clown emerges.

    a. Draw a force diagram for the clown at an arbitrary moment of time during his flight.

    b. Based on Newton's Second Law, describe the clown's acceleration vector during his flight. Which way does it point? Does the direction change over time? Does the magnitude change over time? What *is* the magnitude?

    c. Calculate the x- and y-components of the clown's acceleration vector during his flight. Give numerical answers.

    d. The clown's velocity vector just as he emerges from the cannon has horizontal component $v_{0,x} = +20$ m/s and vertical component $v_{0,y} = +40$ m/s. At what angle above the horizontal is the cannon aimed? How fast is the clown traveling when he emerges from the cannon?

e. Find the velocity components v_x and v_y at the moment when the clown reaches the peak of his trajectory. How fast is the clown traveling at this time?

f. How high above the ground is the clown when he reaches the peak of his trajectory?

g. How far from the cannon will the clown land, assuming that the landing height is the same as the launch height?

5. Going back one last time to the elevator in Focused Problem 5 of Chapter 6 and Focused Problem 2 of Chapter 12 (see also Figures 284 and 315), let's suppose that the elevator starts on the ground floor, speeds up from rest at a constant rate to a speed v_{max}, maintains this speed for a while, then slows at a constant rate until it comes to rest at the top floor. Then the sequence of events is reversed as the elevator returns to the ground floor.

a. Sketch a graph of the net force acting on the elevator as a function of time.

b. Sketch a graph of the force of tension in the elevator cable as a function of time.

c. Suppose that the elevator has mass $m = 500$ kg. Take the gravitational force exerted on the elevator by the earth to be 5,000 N. Take its maximum speed to be $v_{max} = 2$ m/s. Also suppose that both acceleration phases of the motion last for 0.5 s. Use this information to label your graphs from (a) and (b) with numerical values.

d. If the cable were going to break, during which phase(s) would it be most likely to do so?

SOLUTIONS CAN BE FOUND ON PAGE 398

CHAPTER 14

Dynamics

The Two Basic Problems of Dynamics

Newton's Second Law permits us to solve two basic kinds of problems. These problems are in a sense the reverse of one another:

- *Problem Type 1.* By observing the motion of an object, deduce the nature of the forces at work on it.

- *Problem Type 2.* By knowing something about the forces at work on an object, predict its future motion.

In his *Principia,* Sir Isaac Newton formulated and solved both kinds of problems. For example, Newton used Johannes Kepler's observation that the planets move in elliptical orbits to conclude that the gravitational force holding the planets in their orbits weakens with distance from the sun, falling off inversely with the square of the distance. By observing the motions of the planets, Newton could infer the nature of the forces at work on them. This was a problem of Type 1.

Next, taking it as established that the gravitational forces acting on the planets vary inversely with the square of their distance from the sun, Newton proceeded to predict the future course of their motion. This was a problem of Type 2.

All of this, mind you, was in the *same book* in which Newton proposed the Second Law in the first place! His contemporaries could perhaps be forgiven for needing some time to take all of this in.

Oh, by the way, Newton also had to invent calculus in order to solve these problems!

Another Look at the Shopping Cart Problem

Back in Chapter 13 we looked at the motion of a shopping cart (Worked Example 2). The graphs we drew for that problem—reproduced here in Figures 209 and 210—provide a great illustration of how the two kinds of dynamics problems work. Study these figures before reading further.

Finding the Forces from the Motion

Let's try a problem of the first type: finding the forces from the motion.

If you know the trajectory
of a target...

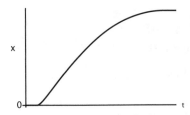

find the rate of change...

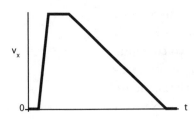

and find it again...

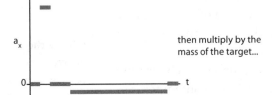

then multiply by the
mass of the target...

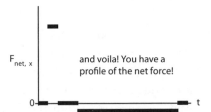

and voila! You have a
profile of the net force!

FIGURE 209. How the first kind of dynamics problem works. Read the words from the top of the page to the bottom, then from left to right.

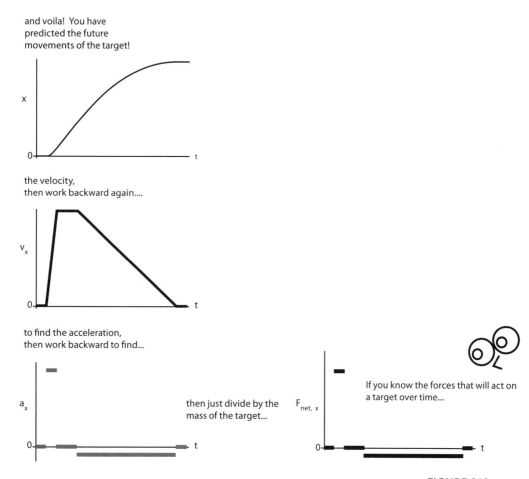

and voila! You have predicted the future movements of the target!

the velocity, then work backward again....

to find the acceleration, then work backward to find...

then just divide by the mass of the target...

If you know the forces that will act on a target over time...

FIGURE 210.
How the second kind of dynamics problem works. Read the words from the lower right part of the page to the left, then from bottom to top.

» WORKED EXAMPLE 1

A 10 g bullet is fired horizontally into a tree trunk at a speed of 500 m/s. Once the bullet enters the tree trunk, frictional forces act over time to bring the bullet to rest in a distance of 4 cm. Find the magnitude and direction of the frictional force that acts to bring the bullet to rest.

» SOLUTION 1

This problem gives us a laundry list of "kinematical" information: data on distances and speeds. We can process this kinematical information to compute the acceleration of the bullet. Then we'll use the acceleration and Newton's Second Law to infer the strength of the friction force responsible for stopping the bullet.

This is one of those situations in which we are not given any time data explicitly. Nor are we asked for any time data. So the most convenient kinematical formula to use is the one without any time variable in it (see Focused Problem 10 of Chapter 9):

(199)
$$v_{f,x}^2 - v_{i,x}^2 = 2a_x \Delta x.$$

A formula like Equation 199 always refers to a chosen coordinate system. (Otherwise, what on earth do all of those "x" subscripts mean?) So before proceeding, we'd better draw a cartoon with a coordinate system laid down over it. I've done that in Figure 211. (How did I know to draw **a** the way I did?)

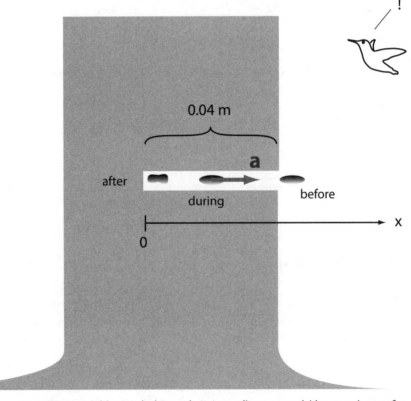

FIGURE 211. Solving Worked Example 1. A coordinate system laid onto an image of a bullet shot into a tree shown at three instants of time: just before striking the tree, during the slowing-down process, and finally at rest.

Looking at the "initial" (before) and "final" (after) situations in Figure 211, we assign the following values as "givens":

$$x_i = +4 \text{ cm} = +0.04 \text{ m} \tag{200}$$

$$x_f = 0 \tag{201}$$

$$v_{i,x} = -500 \text{ m/s} \tag{202}$$

$$v_{f,x} = 0. \tag{203}$$

Inserting these values into Equation 199, we have

$$(0)^2 - (-500 \text{ m/s})^2 = 2a_x(0 - (+0.04 \text{ m})), \tag{204}$$

which is easily solved to yield

$$a_x = +3.13 \times 10^6 \text{ m/s}^2. \tag{205}$$

The plus sign correctly indicates that the acceleration vector points *with* my x-arrowhead, as shown in Figure 211.

The y-component of the acceleration vector is zero, as shown in the figure. This is because we're assuming that the bullet is not turning at all as it embeds itself in the tree trunk.

Is it possible for an actual bullet to turn up, down, left, or right as it embeds itself in a tree trunk? Absolutely. But I'd rather not analyze that situation if I can help it. Or, at any rate, I'd rather analyze the straight-line version first! I like to tell my students this: No problem can possibly list all the hidden assumptions behind it. Your job is to recognize what assumptions *you* are making. And you should make as many assumptions as necessary without making the problem meaningless. As physicists like to say, "Everything should be made as simple as possible, but not simpler."

Because $a_x = +3.13 \times 10^6$ m/s^2 and $a_y = 0$, the direction of the acceleration vector is purely along the x-axis, and the magnitude of the acceleration vector is $a = \sqrt{a_x^2 + a_y^2} = |a_x| = 3.13 \times 10^6$ m/s^2. That's a tremendous braking acceleration!

Once the magnitude of the acceleration during the slow-down phase is known, the magnitude of the net force acting during that phase is easily found:

$$F_{\text{net}} = ma \tag{206}$$

$$= (0.010 \text{ kg})(3.13 \times 10^6 \text{ m/s}^2) \tag{207}$$

$$= 31,300 \text{ N}. \tag{208}$$

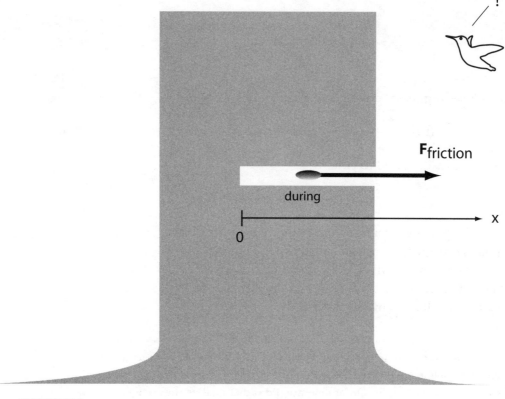

FIGURE 212.
Solving Worked
Example 1. The
friction force from
Figure 211, which is
the only noticeable
force on this scale.

This is a pretty strong force—almost 1,500 pounds! (To convert from newtons to pounds, divide by 21.6.)

This is the *net* force acting on the bullet during the slow-down phase. The problem asks us for the *friction* force acting on the bullet, which is only one of the forces at work. But clearly the friction force acting on the bullet is the dominant force in the problem; all other forces, such as the weight of the bullet, are negligible in comparison. See Figure 212. So the friction force basically *is* the net force. Therefore, our answer regarding the friction force is 31,300 N. ◇

Do you no-
tice how I'm
always check-
ing, checking,
checking things?
When you fin-
ish a problem,
you're not fin-
ished. Mistakes
are inevitable.
Find them!

Notice that according to this force diagram the net force points in the same direction as the acceleration vector of the bullet during the slow-down phase, as required by Newton's Little Law.

To use the formula $v_{f,x}^2 - v_{i,x}^2 = 2a_x\Delta x$, we had to assume that the x-acceleration of the bullet was constant over time. Is this realistic? Is the friction force really constant in strength during the slow-down phase? Almost certainly not. The different layers

FIGURE 213.
A cutaway view of
a bullet fired into a
block of wood.

of the tree trunk have very different material properties. And the wood and the bullet are both being torn apart during the collision, splintering, heating up, and changing shape. See Figure 213, which shows a cutaway view of a bullet fired into a block of soft pine. The tunnel bored by the bullet has a dark coating of lead. This lead was scraped off by the friction force we just calculated in Worked Example 1.

In truth, the force on the bullet from the tree varies wildly during the interaction. My rough intuition about this is that it's something like what is shown in Figure 214. The meaning of our answer "31,300 N" is indicated by the horizontal dashed line. As you can see, our analysis has really just nailed down the general order of magnitude of the net force involved.

Finding the Motion from the Forces

In Worked Example 1 we computed the force acting on an object, in this case a bullet, by using information about the motion of the object. Now we'll do the reverse: describe the future motion of an object by knowing what forces will act on it.

» WORKED EXAMPLE 2

A block of mass $m = 2$ kg is sliding to the right along a frictionless section of ground, moving in a straight line with a constant speed of 5 m/s. At time $t = 0$, the block enters a region where the ground's

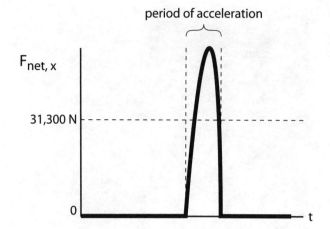

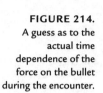

FIGURE 214.
A guess as to the actual time dependence of the force on the bullet during the encounter.

surface is a bit rough. The friction force exerted by the ground has a constant magnitude of 1.96 N. When will the block come to rest, and where will it be when it stops?

» SOLUTION 2

A cartoon of the situation (with a coordinate system) is shown in Figure 215.

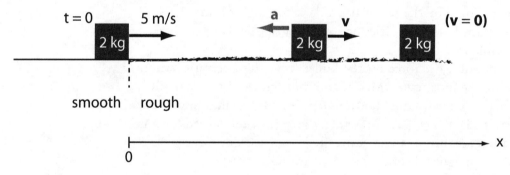

FIGURE 215.
Solving Worked Example 2. Cartoon of a block sliding to the right along a frictionless bit of ground, then encountering a region where the ground is a bit rough.

I have also drawn a force diagram for the block in Figure 216. You can see from the force diagram that there is an unbalanced x-force on the block, so the block will have a nonzero x-acceleration. This acceleration vector is shown on the original cartoon Figure 215.

Looking at the force diagram, we can see that the x-component of the net force is $F_{\text{net},x} = -1.96$ N. From $F_{\text{net},x} = ma_x$, it follows that the horizontal acceleration of the block is $a_x = -1.96\ \text{N}/(2\ \text{kg}) = -0.98\ \text{m/s}^2$.

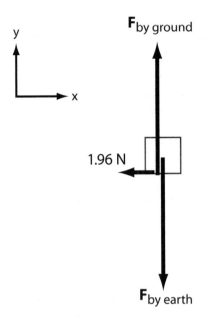

y

x

Fby ground

1.96 N

Fby earth

FIGURE 216.
Worked Example 2.
A force diagram for a
moment in time when
the block seen in
Figure 215 is slowing
down.

What about the y-component of the acceleration vector? Because the velocity vector is horizontal, a vertical component of the acceleration would constitute a turning component $(a_\perp)$. But if the block is truly moving along flat ground, it will not veer up or down, and there will be no turning component. So the y-component of the acceleration vector will always remain zero. By the way, this tells us that the upward force $\mathbf{F}_{\text{by ground}}$ in Figure 216 must exactly balance the downward gravitational force $\mathbf{F}_{\text{by earth}}$, as the diagram suggests. Any imbalance in these forces would cause the block to veer upward into the air or sink downward into the ground.

Altogether then, the acceleration vector is $\mathbf{a} = (-0.98 \text{ m/s}^2, 0)$. Knowing the acceleration vector, we can now say various things about what will happen to the block. For example, the change in the block's position from beginning to end can be found from $v_{f,x}^2 - v_{i,x}^2 = 2a_x \Delta x$:

$$\Delta x = \frac{v_{f,x}^2 - v_{i,x}^2}{2a_x} \tag{209}$$

$$= \frac{(0 \text{ m/s})^2 - (+5 \text{ m/s})^2}{2(-0.98 \text{ m/s}^2)} \tag{210}$$

$$\Delta x = +12.8 \text{ m}. \tag{211}$$

This says that $x_f - x_i = +12.8$ m, but because the initial position was at the origin, we have $x_i = 0$, and therefore $x_f = +12.8$ m. In other words, when the block comes to rest, it will be 12.8 m to the right of the place where the ground becomes rough.

At what time will the block reach its final location? We can find this most easily from the kinematics formula $v_x(t) = v_{0,x} + a_x t$:

(212)
$$v_x(t) = v_{0,x} + a_x t$$

(213)
$$0 = (+5 \text{ m/s}) + (-0.98 \text{ m/s}^2)t,$$

which gives $t = 5.1$ s. Knowing the force that acts, we have been able to determine the future motion of the block—when it will stop and where it will be at that time. ◇

The Deterministic Nature of Newtonian Mechanics

At first it is not so obvious that Newton's Laws spell out a deterministic worldview. But Worked Example 2 (and Figure 210) give a sense of how determinism emerges from the theory, for in that problem we saw that if you know the forces that will act on an object over time, you can determine the acceleration of the object over time. And from the acceleration you can deduce the entire future trajectory of the object before the object actually traces that trajectory. (You do have to know the position of the object at $t = 0$ and the velocity of the object at $t = 0$; see Worked Example 9 in Chapter 9 for a good illustration of this.)

The dynamics problems we have seen in this chapter are simplified in a few ways. For one thing, there has been only one target in each of these problems. But in a more complicated situation—such as calculating how the earth affects the moon's motion (and vice versa)—there will be two (or more!) targets influencing each other simultaneously.

Also, in Worked Example 2, the force of 1.96 N acting on the block was known to be constant in time. In more sophisticated examples (such as the earth-moon situation, for example), the forces are indeed known, but not as explicit functions of time; rather, they are known in terms of the *relative positions* of the sources and targets. In the earth-moon problem, the earth and moon start off with given positions and velocities and, by virtue of their relative positions, they exert forces on each other. These forces cause them to accelerate, which causes them to move to new positions, which causes them to exert new forces on each other. These forces cause them to accelerate in new ways, which causes them to move in new ways, which causes them to exert new forces on each other, and so on!

What's true for the earth and the moon is true for the sun and all the planets at once. And also for all the stars in the Milky Way galaxy—and for all the galaxies in the universe! Each body exerts a force on all the others; these forces alter the bodies' motions, and the changing positions of the bodies cause new forces to be exerted, which changes the bodies' motions in new ways, and so on. It is a complex and beautiful dance—but also one without improvisation; one whose steps are entirely prescribed in advance.

Never has this idea been more beautifully expressed than by Pierre-Simon de Laplace in his *Essai Philosophique des Probabilités*, first published in 1814:

> We ought then to regard the present state of the universe as the effect of its anterior state and as the cause of the one which is to follow. Given for one instant an intelligence which could comprehend all the forces by which nature is animated and the respective situation of the beings who compose it—an intelligence sufficiently vast to submit these data to analysis—it would embrace in the same formula the movements of the greatest bodies of the universe and those of the lightest atom; for it, nothing would be uncertain and the future, as the past, would be present to its eyes [6].

Laplace took the deterministic thesis to its ultimate end. According to Laplace, it is not only the future motions of the planets that have already been determined today, but also the future motions of those "tiny planets" that are the atoms in our own bodies!

The Laplacian vision is deeply challenging to the idea of free will. For if our minds reside in our brains, and if our brains are made of atoms, and if our thoughts in some way correspond to the motions of those atoms, and if those motions are already prescribed for all future time—indeed, if they have been prescribed in advance since the beginning of the universe—am I really in control of my own thoughts and actions?

For that matter, can there be any such thing as a "random" event in Laplace's universe? Imagine a ball on a roulette wheel landing on the number 36. We are accustomed to regard events like this as random. But in the Laplacian view, the ball landed on 36 because of the forces that drove it there, and those forces arose because of the prior position of the ball and its surroundings. And those things came to be where *they* were because of the forces that drove *them* there. That these surroundings include gamblers and a croupier is irrelevant; after all, the crowd around the roulette table is just another constellation of atoms.

Ironically, Newton himself would not have accepted any of this! Newton was a *dualist*. He believed that the universe was made of two kinds of stuff: matter (which always obeys the Second Law of Motion) together with mental objects, such as the soul (which are endowed with free will). According to Newton, the physical corpus, or body, is controlled by the soul, much as a dockworker pilots a crane. Newton's analysis of the roulette wheel would say that the croupier freely decides when to release the ball—meaning that no law of physics could ever predict which number the ball will land on.

Today our fundamental theories of nature are *non*deterministic. So a physicist of our time can believe in chance events without having to be a dualist. But what about the question of free will? Personally, I'm not sure what, if anything, physics has to tell us— or has ever had to tell us—about that.

Reader's Notes

Below is space for you to summarize key points from this chapter.

FOCUSED PROBLEMS :: CHAPTER 14

1. Figure 217 shows the x-coordinate of a person pacing the floor. Sketch a graph of the net force $F_{net,x}$ acting on the person during this period of time. What object is exerting this force on the person?

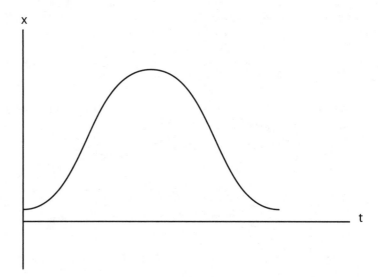

FIGURE 217.
Focused Problem 1.
The x-coordinate of a
person pacing the
floor.

2. Suppose that the net horizontal force on a trolley car during the interval $0 \leq t \leq 5$ s is known to be of the form $F_{net,x}(t) = -(20{,}000 \text{ N/s})\, t$. Sketch $F_{net,x}$ as a function of time. From this, sketch the position x of the trolley car as a function of time, assuming that $x = 0$ at $t = 0$ and $v_x = 10$ m/s at $t = 0$. Assume that the motion continues until the car comes to rest. What would it feel like to ride on this trolley car during the period in question?

3. [Uses calculus.] Give explicit formulas for $x(t)$ and $v_x(t)$ for the trolley car in Focused Problem 2. Let the mass of the car be $M = 25{,}000$ kg. How far does the trolley car travel from $t = 0$ until the time it comes to rest?

The next three problems involve projectiles. I recommend revisiting Focused Problem 4 back in Chapter 13 before doing these problems.

---

4. A pelican with a fish in its mouth is flying horizontally due west at 5 m/s, at an altitude 20 m above sea level. At time $t = 0$, it passes directly over a sea otter floating on the surface and lets go of its fish at that instant.

   Assume that gravity is the only important force acting on the fish during its fall. Take the acceleration vector of the fish as having the constant value $\mathbf{a} = (a_x, a_y) = (0, -9.8 \text{ m/s}^2)$, where the positive y direction is upward.

   a. How far from the sea otter will the fish strike the water?
   b. When the fish strikes the water, what will its velocity be? (Give the magnitude and direction of the velocity vector.)

5. Poindexter is throwing a ball against a wall. He throws the ball upward with an initial velocity of 10 m/s at an angle 30° above the horizontal. The release point is 1.5 m above the ground. The wall is 10 m away.

   Assume that gravity is the only important force acting on the ball during its flight. Take the acceleration vector of the ball as having the constant value $\mathbf{a} = (a_x, a_y) = (0, -9.8 \text{ m/s}^2)$, where the positive y direction is upward.

   a. How high up the wall will the ball strike?
   b. When the ball strikes the wall, what will its velocity be? (Give the magnitude and direction of the velocity vector.) Will the ball be on the way up or on the way down when it hits?

6. The Indonesian volcano Krakatoa exploded on August 27, 1883, in the fifth-largest explosion known in the geologic record [7]. The sound was heard by people 3,000 miles away. Rocks were catapulted to a height of 20 miles or more. Assuming an initial launch angle of 45°, what was the speed of these ejected rocks?

   Assume that gravity was the only important force acting on the rocks while they were in flight. Take the acceleration vector of the rocks as having had the constant value $\mathbf{a} = (a_x, a_y) = (0, -9.8 \text{ m/s}^2)$, where the positive y direction was upward.

7. [Uses calculus.] Suppose that the net force acting on an object of mass m is known as a function of the object's position as

$F_{net,x} = -kx$, where x is the horizontal position and $k > 0$ is a constant. This is an example of a *restoring force,* because the force points toward the origin at all times, tending to keep the target in the vicinity of the origin. This is the force law we would use if we were analyzing the 1-D motion of a mass on a spring (see Chapter 16).

a. Write out the Second Law for this situation in the form of a differential equation in $x(t)$.

b. Suppose that the target's initial position is x_0 and its initial horizontal velocity is $v_{0,x}$. Show that the trajectory $x(t) = x_0 \cos \omega t + \frac{v_{0,x}}{\omega} \sin \omega t$ satisfies the differential equation as well as the initial conditions, provided the constant ω is chosen properly. What must ω be in terms of the parameters of the problem?

c. Sketch numerically accurate $x - t$, $v_x - t$, $a_x - t$, and $F_{net,x}$ graphs for the target, assuming that it starts from rest at position $x_0 = +1$ m. Take $k = 10$ N/m and $m = 0.25$ kg.

SOLUTIONS CAN BE FOUND ON PAGE 402

Newton's Third Law

In Chapter 10 I explained that every action is an *inter*action. You can't touch without *being* touched. Newton's Third Law formalizes this idea and makes it quantitative. Here it is:

> *Newton's Third Law:* If object 1 exerts a force on object 2, object 2 must also exert an equal and opposite force on object 1.

This is the famous law of "action-reaction." It says that if you exert a force on me, I also exert a force on you; and it says that this "reaction force" is equal in magnitude but opposite in direction to the force you exert on me. Figure 218 shows the sun's pull on the earth and the earth's pull on the sun, which are equal and opposite to one another. Of course, the resulting accelerations of the two bodies are very different owing to their great disparity in mass.

Third Law Pairs

The two forces shown in Figure 218 are known as a "Third Law pair." The two members of a Third Law pair are the two faces of a direct and mutual interaction.

If Alice and Bob lean on each other as shown in Figure 219, Bob's push on Alice and Alice's push on Bob are Third Law pairs. These forces are necessarily equal in strength and opposite in direction.

But if Alice and Bob pull against each other with a rope between them, as shown in Figure 220, they are actually no longer exerting

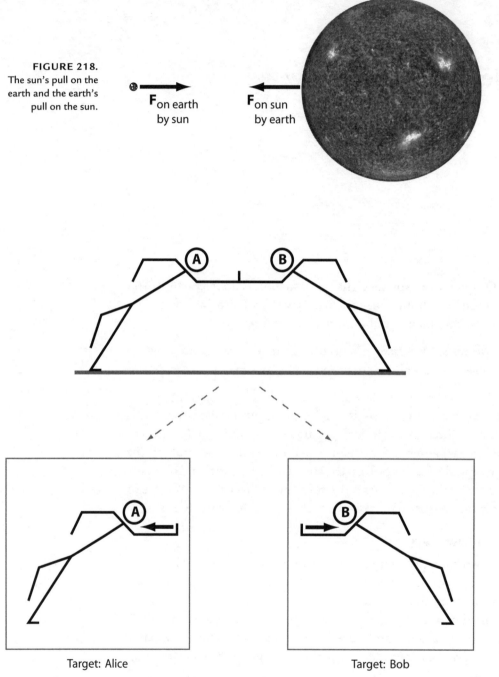

FIGURE 218.
The sun's pull on the earth and the earth's pull on the sun.

$F_{\text{on earth by sun}}$

$F_{\text{on sun by earth}}$

Target: Alice

Target: Bob

FIGURE 219. Alice and Bob leaning against one another, exerting Third-Law-pair forces on each other. Only the forces of direct contact between them are shown.

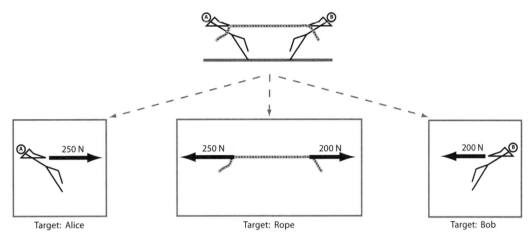

FIGURE 220. Alice and Bob pulling a rope, no longer exerting forces on each other.

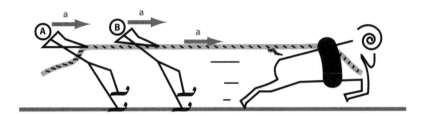

FIGURE 221. Worked Example 1. Alice and Bob being towed by a ram by means of a rope.

forces on each other at all—because they aren't in direct contact! They are both exerting forces on the rope, and the rope is exerting forces on each of them.

Because Alice and Bob are no longer in direct contact, there is no reason that the force on Alice must have the same magnitude as the force on Bob. The Third Law applies only to forces of direct interaction.

Note that the diagrams shown in Figure 220 are not complete force diagrams; no gravitational forces are shown, for example.

When multiple bodies are interacting with one another, as in Figure 220, don't try to think about all of them at once. Divide and conquer; take it one target at a time.

» WORKED EXAMPLE 1

Alice and Bob, on ice skates, are holding onto a rope that is fastened to a crazed ram and being towed by it. See Figure 221. Alice, Bob, the rope, and the ram are all accelerating at 1 m/s^2 to the right.

Draw rough non-numerical but plausible force diagrams for Alice, Bob, and the rope. Identify all of the Third Law pairs.

For simplicity, ignore any vertical forces, as well as any friction forces and forces of air resistance.

» SOLUTION 1

When I say "plausible," I mean that the net force vector on each target should resemble the acceleration vector of each target. My force diagrams are shown in Figure 222.

Target: Alice

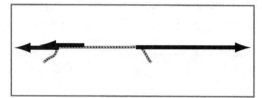

Target: Bob   Target: Rope

FIGURE 222.
Solution to Worked Example 1. Force diagrams for Alice, Bob, and the rope.

Notice that for each target, the net force on the target points to the right, matching the target's acceleration vector. Only horizontal forces are shown. Also, the force on the rope from Bob is shown displaced slightly upward so that you can see the vectors better. ◇

Can you identify the Third Law pairs in Figure 222? In Figure 223, I have numbered the forces 1 through 5. The Third Law pairs are (1, 3) and (2, 4). Vector 5 is not part of a Third Law pair; to see its Third Law counterpart, we'd have to draw a force diagram for the ram.

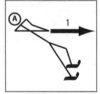

Target: Alice

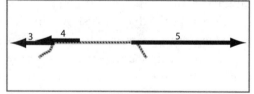

Target: Bob   Target: Rope

FIGURE 223.
Solution to Worked Example 1. Numbering the forces from Figure 222.

» WORKED EXAMPLE 2

Imagine a huge cement truck smashing head-on into a speeding motorcycle. Many people find it hard to believe that during the collision the motorcycle exerts just as large a force on the truck as the truck exerts on the motorcycle. They tend to think that the truck exerts a larger force. Why do you think people reason this way? What is the resolution of the paradox?

» SOLUTION 2

In a collision between a cement truck and a motorcycle, the motorcycle will suffer violent consequences, whereas the cement truck will hardly be affected at all. But this is perfectly consistent with the forces on each object being the same. For example, if the cement truck has a mass of 10,000 kg and the motorcycle has a mass of 200 kg, during a collision in which each exerts a force of 10,000 N on the other, the cement truck will have an acceleration on the order of $10{,}000\,\text{N}/10{,}000\,\text{kg} = 1\,\text{m/s}^2$, but the motorcycle will have an acceleration on the order of $10{,}000\,\text{N}/200\,\text{kg} = 50\,\text{m/s}^2$. The cement truck will barely slow down, but a few instants after the collision the motorcycle will violently reverse direction entirely. See Figure 224. Only the forces of interaction are shown; the vertical forces in this situation largely cancel, and other horizontal forces are negligible compared to the interaction forces.

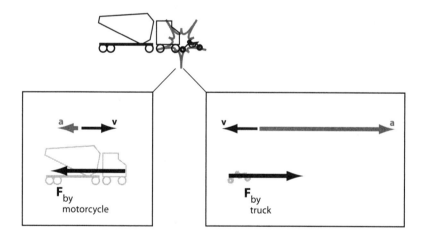

FIGURE 224.
A moment of time during a collision between a cement truck and a motorcycle.

I suspect people believe the truck exerts a larger force for two reasons. First, the motorcycle suffers the greater consequences, so they deduce (incorrectly) that it experienced the greater force. And second, I suspect that when people imagine this situation they are actually not even thinking about force (interaction) at all but rather about momentum—about oomph. They are thinking of something havable. And if it's something havable we're talking about, then sure, I agree; the cement truck is going to have lots more of it. ◇

Reader's Notes

Below is space for you to summarize key points from this chapter.

FOCUSED PROBLEMS :: CHAPTER 15

1. *The lazy horse; or, a little bit of knowledge is dangerous.* This is an old paradox. The lazy horse says, "Why should I even try to pull a cart? By Newton's Third Law, the cart is just going to pull back on me with an equal force, so I won't be able to move forward anyway." Of course, in real life we know that the horse can speed up from rest if he wants to. Can you see how this comes about?

2. *The bully.* A big bully comes up behind a little nerd and gives him a shove. The little nerd falls over. If the nerd exerted just as large a force on the bully as the bully exerted on the nerd, why didn't they *both* fall?

3. *Revenge of the nerd.* Next day, the nerd sneaks up on the bully and gives *him* a shove. The bully falls down; the nerd doesn't. How is *that* possible?

4. *Monkey in a rocket.* Our friend the swinging orangutan from Chapter 10 is back, this time on a spaceship to Mars. See Figure 225.

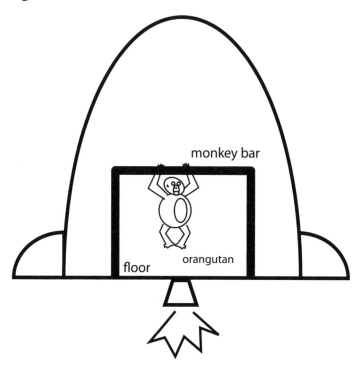

FIGURE 225. Focused Problem 4. An orangutan hanging from a monkey bar inside a rocket.

The spaceship is accelerating in the forward direction (upward in the figure). Sketch plausible force diagrams for the orangutan and the monkey bar. Identify all of the Third Law pairs. (Neglect any gravitational forces.)

5. *Hummingbird trapped in a box.* A hummingbird is flying near the top of an airtight box, using her beak to press upward on the underside of the box top. If the hummingbird were strong enough and the box were light enough, could the box be lifted this way?

SOLUTIONS CAN BE FOUND ON PAGE 405

CHAPTER 16

Kinds of Force

Every force has a *type,* or *kind.* For example, the force that the sun exerts on the earth is of the gravitational kind. The force that a magnet exerts on a nail is of the magnetic kind. The force that lifts your hairs after you rub a balloon on your head is of the electrostatic kind. Table 5 lists the kinds of force that we commonly see in Newtonian physics.

Not all of these categories of force are what we'd call *fundamental.* For example, friction forces are ultimately electrostatic in nature. But for the purposes of this book, it is not always necessary or even useful to reduce forces to their fundamentals. (If the goal is to understand the motion of an airplane, a planet, or a dancer, you won't get very far using quantum theory.)

This chapter summarizes the basics of most of the kinds of forces you will encounter in your study of Newtonian mechanics.

Two Different Questions

Imagine tossing a ball into the air. As we all know, the ball eventually comes down. Question: What makes the ball come down?

I can think of two very reasonable answers to this question. One answer: *gravity.* Gravity is what made the ball come down.

But there is another reasonable answer to the question: namely, *the earth.* The earth, after all, is the *material object* that is *doing something* to the ball to make it come back down.

So when we ask, "What makes the ball come down?" we should really be more precise about what we are asking. To summarize by way of questions and answers:

Question: What *object* is influencing the ball so that it comes down?

Answer: The earth.

Question: What *kind of force* is influencing the ball so that it comes down?

Answer: A gravitational force.

In the rest of this chapter we'll summarize the basic behaviors of some of the most useful kinds of force in Table 5: gravitational forces, "normal" forces, frictional forces, tension forces, spring forces, and drag forces.

Gravitational Forces

I'm going to summarize how we calculate the gravitational force acting on a target object. Actually, I'll give three minisummaries: one for the special case when the target object is located near the surface of the earth, one for the special case when the target object is located near the surface of some other planet, and finally, a summary of Newton's general theory, which covers all cases.

Target located near the earth's surface. If a target object of mass m is located near the earth's surface, the earth will exert a force on the target of magnitude mg. Succinctly,

(214)
$$F_{\text{by earth}} = mg.$$

Here, g is a measured constant with the value $g \equiv 9.8 \text{ N/kg}$. The value 9.8 N/kg represents the intrinsic strength of the earth's gravitational field at the planetary surface. The direction of the force is toward the center of the earth.

You may wonder just how near the earth's surface the target has to be in order to use the formula $F_{\text{by earth}} = mg$. Well, it depends on how accurate you want to be. If you are within the earth's atmosphere, $F_{\text{by earth}} = mg$ is very accurate for most purposes.

Target located near the surface of some other moon, planet, or similar object. If the target mass m is located near the surface of our moon, the moon will exert a gravitational force on the target of a magnitude equal to $F_{\text{by moon}} = mg_{\text{moon}}$, where $g_{\text{moon}} = 1.6 \text{ N/kg}$. The moon's gravitational field is weaker than the earth's. The direction of the force will be toward the center of the moon.

The story for other celestial bodies is the same. Each celestial body has its own surface-gravity constant, g_{body}, and the gravita-

In this book, whenever we talk about the "weight" of an object, we are actually referring to the gravitational force exerted on the object by the earth.

TABLE 5. Kinds of Force Commonly Encountered in Newtonian Physics and the Material Objects That Produce Them

Kind of force	Source objects giving rise to the force
Gravitational	Massive objects
Electrostatic	Charged objects
Magnetic	Magnets, current-carrying wires
"Normal"	Hard surfaces
Frictional	Rough surfaces
Tension	Ropes and strings, rods
Spring (restorative)	Springs
Drag	Fluids in motion or in relative motion
Lift	Fluids moving past a wing
Thrust	Flapping wings, paddling arms, exhaust gases from an engine, etc.

tional force exerted on a target m by the celestial body will be given by

$$F_{\text{by body}} = mg_{\text{body}}. \tag{215}$$

If the celestial body is pretty round, the direction of the force will be toward the center of the body.

Universal gravitation. The previous two scenarios were special cases of a general rule. That general rule is Newton's law of universal gravitation. This law states that any two objects in the universe exert attractive gravitational forces on each other. The forces are equal and opposite. The forces are weaker if the two objects are farther apart. Specifically, given two pointlike objects of mass m_1 and m_2 separated by a distance r, we have

$$F_{\text{on 1 by 2}} = F_{\text{on 2 by 1}} = \frac{Gm_1m_2}{r^2}. \tag{216}$$

Here G is a universal constant with the measured value $G = 6.67 \times 10^{-11} \, \text{m}^3/\text{kg} \cdot \text{s}^2$.

If an object is spherically symmetric—or, like a planet, approximately so—you can use the pointlike formula $F_{\text{on 1 by 2}} = \frac{Gm_1m_2}{r^2}$ to find the gravitational force that the object exerts or experiences.

Just consider all the mass of the object to be concentrated at its center. (It takes calculus to show why this trick works; Newton himself delayed the publication of his *Principia* for years until he'd solved this problem to his satisfaction.)

» WORKED EXAMPLE 1

An astronaut of mass 100 kilograms is in a spaceship orbiting 125 miles above the surface of the earth (Figure 226). Calculate the weight of the astronaut.

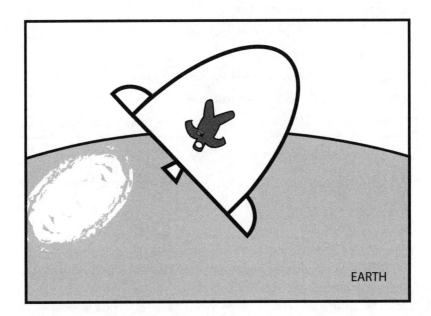

FIGURE 226.
Worked Example 1.
An astronaut
orbiting the earth.

EARTH

» SOLUTION 1

Remember that the "weight" of the astronaut refers to the strength of the gravitational force exerted on the astronaut by the earth. How do we calculate this force?

Astronauts orbiting the earth are really not very far away. Space is closer than you think. In fact, you may be closer to an astronaut right now than you are to some of your brothers or sisters. So if we consider the astronaut to be located "near" the surface of the earth, we can find his weight using the simple formula $F_{\text{by earth}} = mg$. This gives

$$F_{\text{by earth}} = (100 \text{ kg})(9.8 \text{ N/kg}) \tag{217}$$

$$= 980 \text{ N}. \tag{218}$$

But if we are suspicious of the idea that an astronaut in orbit can safely be considered to be located "near" the earth, we have to use the universal gravitation formula. The required data are shown in Figure 227.

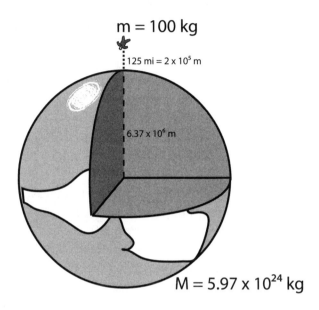

m = 100 kg

125 mi = 2 x 10⁵ m

6.37 x 10⁶ m

M = 5.97 x 10²⁴ kg

FIGURE 227.
Solving Worked Example 1. A cutaway view of the earth showing inputs for the gravitation formula. Distances are not to scale.

Using these values in the universal formula, we have

$$F_{\text{by earth}} = \frac{Gm_1m_2}{r^2} \tag{219}$$

$$= \frac{\left(6.67 \times 10^{-11}\frac{\text{m}^3}{\text{kg·s}^2}\right)(100\text{ kg})(5.97 \times 10^{24}\text{ kg})}{(6.37 \times 10^6\text{ m} + 2.0 \times 10^5\text{ m})^2} \tag{220}$$

$$= \frac{\left(6.67 \times 10^{-11}\frac{\text{m}^3}{\text{kg·s}^2}\right)(100\text{ kg})(5.97 \times 10^{24}\text{ kg})}{(6.57 \times 10^6\text{ m})^2} \tag{221}$$

$$= 923\text{ N}. \tag{222}$$

So the astronaut's true weight is more like 923 N, not 980 N. By using the simple "mg" formula, we overestimated the weight of the astronaut by 57 N, or about 6%. Does 6% matter? If all you want is a gut feeling for the weight of an astronaut, no. But if you want to get him home safely, 6% matters a lot! ◇

Many people are surprised to learn that an astronaut orbiting the earth is almost as heavy (923 N) as an astronaut standing on terra firma (980 N). Aren't astronauts "weightless"?

No. Astronauts are not weightless. The earth's gravitational field extends far out into space. The gravitational field doesn't magically shut off as soon as you get outside the atmosphere.

The Feeling of "Weightlessness"

But surely *something* funny is going on here, because the astronaut certainly *feels* weightless, doesn't he?

This question represents a conflict of intuitions that gives students no end of trouble, so let me take a minute to talk about it.

In orbit, the astronaut is subject to a force of 923 N—almost as strong as the force of 980 N that he would be subject to at sea level. Yet in space he feels weightless. How can he feel so different in the two situations when the gravitational force on him is so nearly the same in each?

First of all, we should note that an orbiting astronaut is accelerating—moving in a circular orbit, say. Because the astronaut is experiencing life in an accelerating frame of reference, he or she (and we) should not even try to analyze his or her feelings in terms of forces and the Second Law. (See the section headed "Accelerating Frames of Reference: A Caution" in Chapter 13.)

Second, let me point out that there is nothing particularly special about "being in orbit." You don't have to become an astronaut to go into orbit. When you leap toward the hoop in basketball, you're going into orbit. When you're jogging, during the moments when your feet are not touching the ground, you're in orbit. The difference between an astronaut and a long-jumper is only a difference of degree.

Third, let me point out that no one can sense gravitational fields directly. All we know is what the nerves in our bodies tell us. The nerves in your body don't sense gravitation. They sense pressure or tension. In short, they sense mechanical stresses within your tissues.

With this in mind, consider Figure 228, which shows the forces acting on a person standing on the ground.

When you're standing on the ground, there are two kinds of force acting on you: the gravitational force from the earth pulling you down and forces from the ground pushing you up. This means that you're effectively standing in a vise! This is easier to see if you tilt your head and look at the diagram sideways. See Figure 229, where I've moved the point of application of the gravitational force to the person's head, just to make a point.

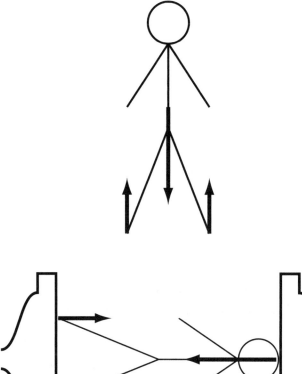

FIGURE 228.
Force diagram for
a person standing on
the ground.

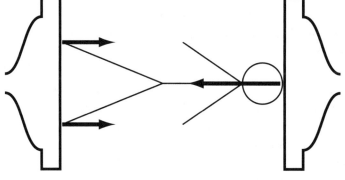

FIGURE 229.
Figure 228 viewed
sideways.

When you're in a vise like this, your tissues are compressed.
Your vertebrae squeeze together. Your soft organs flatten out. This
flattened state is monitored by your nerves. And when you change
your state—for example, by lying down—your guts redistribute
themselves and your nerves communicate the new configuration
to your brain. In response, your brain says, "Ahhhh."

What happens when you go into orbit—say, by stepping off a
cliff? See Figure 230. Now you're no longer touching the ground, so
there is only one force acting on you: the gravitational force. Just
like that, the vise is gone! Your tissues expand, and your nerves
communicate the new state of affairs to your brain. Your brain
interprets this as weightlessness. But there's still plenty of weight
here, in the sense that the force exerted on you by the earth has not
changed. What you're really feeling is not so much weightlessness
as it is "viselessness"!

FIGURE 230.
A person in free fall.

So an astronaut feels weightless not because he or she has no weight but because his or her body is not being compressed by opposing forces.

I think of those cartoons where a hunter falls off a cliff. Have you ever seen one of those? In the cartoons, the camera always falls alongside the hunter, tracking his fall. For us at home watching the cartoon, we see the hunter floating motionless within the frame. Perhaps the hunter has a cap that lifts gently and floats a few inches above his head. Doesn't this movie look almost as though it could have been filmed inside of an orbiting space station? Well, it should look that way! The physics of the two situations is entirely the same in the sense that we are watching the effects of gravity unopposed by any walls or floors. As long as the hunter (or the astronaut) is subject only to gravity, he or she will enjoy the feeling of viselessness. For the hunter, the sensation will be short-lived!

This is all I'm going to say about gravitational forces in this chapter. Let's move on to the other important kinds of force.

"Normal" Forces

You're walking across a room, not paying attention, when suddenly you bump into a wall. During the time that you are in contact with the wall, the wall is exerting a force on you. This force varies in strength throughout the interaction; it disappears entirely when you finally break contact with the wall and back up. At all times,

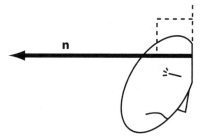

FIGURE 231.
A normal force.

the force exerted on you by the wall is directed straight away from the surface of the wall.

This is an example of a "normal" force. The word *normal* here is actually a bit of mathematical jargon; *normal* is a math word meaning "perpendicular." Normal forces are called normal because they point in a direction perpendicular to the surface that creates them. See Figure 231, which shows a person's head bumping into a wall. The force exerted on the head by the wall is perpendicular to the wall; hence the name "normal" force.

Whenever a hard object presses against another hard object, the two objects will resist interpenetrating one another by exerting normal forces to repel one another. For example, if somebody punches you, his hand exerts a normal force on your head during the period of contact. (At the same time, your head also exerts a normal force on his hand. He may even suffer some broken hand bones because of it.)

When a fastball heading toward home plate suddenly reverses direction and sails over the fence, the violent reversal in its motion is caused by the stupendous normal force that is exerted on the ball by the bat during the brief moment of contact between bat and ball.

Earlier I gave several formulas for calculating the strength of gravitational forces, including $F_{\text{by earth}} = mg$ and $F_{\text{on 1 by 2}} = Gm_1m_2/r^2$. Sometimes students ask me what the formula for the normal force is. The answer is that there is no formula. Normal forces are a dynamic response to a stimulus: You push the wall, the wall pushes back. You push the wall harder, the wall pushes back harder. This responsive quality means that there can be no stock formula for the strength of a normal force. It will depend on the particular circumstances.

What a Bathroom Scale Is Telling You

As I said earlier, your "weight" refers to the strength of the earth's gravitational pull on you. Assuming that your mass is 80 kilograms, anywhere near the earth's surface your weight will be $mg = 784$ N. You probably assume that when you stand on a bathroom scale, the scale registers your weight. But what the scale actually registers is how hard your feet push against the scale. This is a normal force, not a gravitational force, and the reading of the scale is not necessarily equal to your weight. Just jump up and down on the scale, and you'll see what I mean! Your weight isn't changing when you jump—but the scale reading certainly is.

Frictional Forces

What keeps the frog in Figure 232 from slipping down the ramp? Aha! This is another of those vague questions, isn't it? Let's be more precise: What *object* keeps the frog from slipping down the ramp? Answer: The *ramp* is the object that keeps the frog from slipping down the ramp. More precisely, the ramp (source) exerts some kind of uphill force on the frog (target) that stops the frog from sliding down the ramp.

The forces acting on the frog are shown in Figure 233. The uphill force exerted by the ramp is labeled $f_{\text{by ramp}}$. You can verify that $F_{\text{net on frog}} = 0$ by adding the vectors tip-to-tail.

Now for another question: *What kind of force* keeps the frog from slipping down the ramp? Answer: a frictional force.

A frictional force is a force that occurs at the interface of two rough surfaces—in this case, the surface of the ramp and the underside of the frog's feet.

FIGURE 232.
A frog on a ramp.

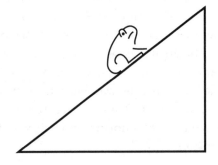

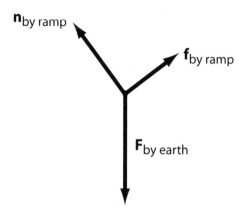

FIGURE 233.
Forces acting on the
frog in Figure 232.

Friction Forces versus Normal Forces

Frictional forces have something in common with normal forces: Both kinds of force occur at an interface where two surfaces touch. But in another respect, frictional forces are very unlike normal forces: Frictional forces always point *along* the surface interface, whereas normal forces always point *perpendicular to* the surface interface.

We can see both friction forces and normal forces at work in Figure 233. We can also see both kinds of force at work in the case of the leaning broom in Figure 167 back in Chapter 10. Look at the point where the broom touches the floor. The force holding up the broom is a normal force exerted by the floor. The force preventing the broom from sliding along the floor is a frictional force exerted by the floor.

Static Friction and Kinetic Friction

Friction forces come in two kinds: *static friction* and *kinetic friction*. Static friction occurs when the two surfaces in question do not slip or move against each other. The force exerted by the ramp on the frog in Figure 232 is a static friction force, because the surface of the ramp and the underside of the frog's feet are not slipping against each other.

Kinetic friction occurs when the two surfaces in question *are* slipping against each other. When you fall on the ground and skin your knee, the force that tears your skin is the force of kinetic friction exerted on your knee as it slides along the ground.

A Formula for Kinetic Friction

There is an easy formula for the strength of a kinetic friction force:

(223)
$$f_k = \mu_k n.$$

(I like to use lowercase fs for friction forces.)

In this formula, n is the magnitude of the normal force of contact between the two surfaces, and μ_k is a dimensionless constant called the "coefficient of kinetic friction." The coefficient of kinetic friction depends on the materials of which the two surfaces are made. Small values of μ_k (roughly 0.01–0.1) indicate a slippery interface, such as that between brass and ice. Large values of μ_k (roughly 0.5–1.0 or larger) indicate a rough interface, such as that between sandpaper and rubber.

According to Equation 223, the strength of the kinetic friction force f_k scales with the normal force of contact between the two surfaces. Intuitively, this is because the harder you press two surfaces together, the stronger the friction between them can become.

The equation $f_k = \mu_k n$ tells us how to calculate the *strength* of a kinetic friction force. The *direction* of the kinetic friction force has to be found by common sense. The friction force opposes the slippage. Figure 234 shows the direction of the kinetic friction forces on your upper teeth and your toothbrush at a particular moment of time when you're brushing your teeth. These two forces are a Third Law pair.

When you press hard to scrub a frying pan, you're increasing the normal force on the pan, which increases the kinetic friction force according to Equation 223. The kinetic friction force then becomes strong enough to dislodge those bits of fried egg.

The Lack of a Formula for Static Friction

There is no formula for calculating the strength of a *static* friction force. It's easy to see why. For example, consider a patient lying etherized on a table (Figure 235). Observe that there is a normal force **n** pushing up on the patient that prevents the patient from sinking into the wood. Observe also that there is *no* friction force acting on the patient. (Nothing is trying to push the patient sideways.)

So far so good. Now suppose that you're pressing lightly forward on the patient with your finger, as shown in Figure 236. If you push lightly enough, the patient won't slip. A static friction force will arise to oppose your push and hold the patient in place. This static friction force points to the left in Figure 236(b).

In the first scenario with the patient (Figure 235), there was no static friction. In the second scenario (Figure 236), you pushed

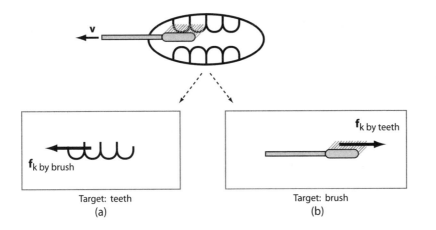

FIGURE 234.
The kinetic friction forces acting on (a) your upper teeth and (b) your toothbrush when you're brushing your teeth.

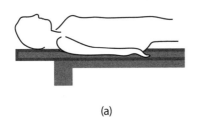

Target: teeth
(a)

$\mathbf{f}_{k \text{ by teeth}}$

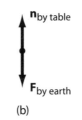

Target: brush
(b)

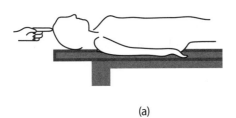

(a)

$\mathbf{n}_{\text{by table}}$

$\mathbf{F}_{\text{by earth}}$

(b)

FIGURE 235.
(a) A patient lying on a table. (b) Forces acting on the patient.

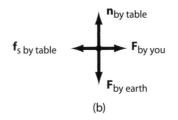

(a)

$\mathbf{n}_{\text{by table}}$

$\mathbf{f}_{s \text{ by table}}$ — $\mathbf{F}_{\text{by you}}$

$\mathbf{F}_{\text{by earth}}$

(b)

FIGURE 236.
(a) Pushing lightly on the patient in Figure 235. (b) The resulting forces acting on the patient.

a little bit, and there was some static friction in response. If you press a little bit harder, there will be a little more static friction. So the static friction force $\mathbf{f}_s$ responds dynamically to the conditions involved and can take a range of values, even while the normal force $\mathbf{n}$ maintains the same strength at all times.

Conclusion: The strength of the static friction force f_s is simply not determined by the value of n. In this respect, static friction is different in an important way from kinetic friction, whose uniform law is expressed in Equation 223.

Static friction takes whatever value it has to in order to keep a surface from slipping in the face of whatever other forces may be at work.

»WORKED EXAMPLE 2

Consider a car driving on a flat, straight road at a constant speed. Draw a plausible force diagram for the car, including the effects of air resistance.

»SOLUTION 2

Because the car is driving on a straight, flat road at a constant speed, its acceleration is zero. Therefore, the net force on the car must be zero. My force diagram is shown in Figure 237. You can check to see that the forces on the car completely cancel each other and add to zero.

Students are often confused by two aspects of the friction force in this problem: (1) the fact that the friction involved is *static* friction, not kinetic friction, even though the car is moving along the road surface, and (2) the fact that the friction force points *with* the car's direction of motion, not against it.

First let's consider my identification of the friction force as *static* friction. Isn't the car moving? So isn't this kinetic friction?

The answer is no. Kinetic friction isn't about moving per se; it's about *sliding*, about the dragging of two surfaces against each other. This car is not spinning its wheels, so there is no sliding or dragging of rubber against asphalt. With no slippage occurring, the friction force at work must be static friction.

This is the same thing that happens when you're walking. When you walk on a surface with good traction, such as dry pavement, your feet don't slide or slip. So any horizontal forces exerted on your feet by the pavement are static friction forces. The sole of your shoe "rolling" along the ground when you're taking a step

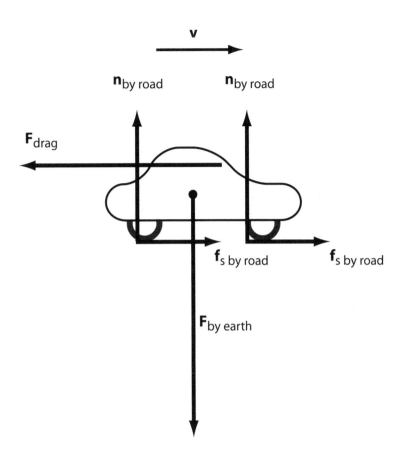

FIGURE 237.
Solving Worked Example 2. Forces acting on a car driving in a straight line at a constant speed. (The car is moving to the right.)

is very like the rolling of a tire along the ground when driving. Figure 238 shows what I mean.

Now let's talk about the direction of the friction force. Students almost never draw the friction force pointing in the proper direction in this problem. I think there are three reasons for this. The first reason is that they have somehow gotten the idea that friction opposes the motion. But friction doesn't necessarily oppose

FIGURE 238.
Illustration of the similarity between the static friction forces that allow you to speed up when you walk and those that allow your car to pick up speed or merely overcome the force of air resistance.

the motion. It opposes *slippage*. Under normal driving conditions, your tires aren't actually slipping against the asphalt. Perhaps you can see that if the tires were to slip, they would skid *backward* along the ground. *That's* why the friction force points forward: It opposes the slip that *could* happen.

The second reason students draw the static friction vectors in the wrong direction is that they may be confusing sources and targets. They are not focusing on the car as the target. They have it in mind that the car pushes backward on the road—which, of course, is true—but somehow don't translate that properly into the road pushing forward on the car (according to Newton's Third Law).

The third reason students draw the static friction vectors in the wrong direction is that they are not thinking about Newton's Second Law. In this particular example, we are told that the car is not speeding up, slowing down, or turning. In other words, we know that $\mathbf{a} = \mathbf{0}$. Therefore, the forces acting on the car must all balance out to zero. But how could they balance out to zero if not only the drag force but also the friction forces pointed backward?

I can put this more simply. We know that the drag force points backward. So if the car is not going to slow down, *something* must be pushing the car forward to compensate. That something is the road. The road pushes the car forward. (After all, what else could be pushing the car?) ◇

Static friction opposes the incipient slip.

The Breaking Point

One thing we've all noticed about friction is that you can "unstick" things if you push hard enough. In the example of the patient in Figure 236, it is clear that if you push hard enough, there will come a moment when the patient does begin to slide across the table. At this point, the friction force will turn into kinetic friction, and Equation 223 will apply. Or again, in the example of the frog in Figure 232, if we imagine making the incline steeper and steeper, there will come a moment when the frog no longer sticks to the ramp but begins sliding downward. At this point, the friction force will turn into kinetic friction, and Equation 223 will apply.

Although static friction takes whatever value it has to in order to keep a surface from slipping in the face of whatever other forces may be at work, it is possible to ask too much of static friction. If you ask too much, then the static friction breaks, the object begins to move, and the friction force turns into kinetic friction described by Equation 223. This is what happens when you unstick the lid on a jar of olives. We call the point at which this happens the breaking point.

Earlier we heard the bad news: There is no formula for the strength of a static friction force *in general*. Now I can give you the good news: There *is* a formula for the strength of a static friction force *right at the breaking point*. A static friction force will break, and the object held in place by that force will begin to slip, as soon as the strength of the static force exceeds a certain limit, which we'll call f_s^{break}. And the formula for f_s^{break} is

$$f_s^{\text{break}} = \mu_s n. \tag{224}$$

In this formula, n is the magnitude of the normal force of contact between the two surfaces, and μ_s is a dimensionless constant called the "coefficient of static friction." As was the case for the coefficient of kinetic friction μ_k, the coefficient of static friction μ_s also depends on the materials of which the two surfaces are made.

Tension

Figure 239 shows a bucket of water being held just above the ground. There must be an upward force on the bucket to prevent it from falling. What object exerts this upward force on the bucket?

Here's the answer: The rope. It is the rope that exerts an upward force on the bucket, holding it up against gravity.

I hope you didn't say that the *person* holding the rope exerts an upward force on the bucket. How can the person do that when the person isn't even touching the bucket?

The kind of force exerted by the rope is a *tension force*. Basically, a tension force is any kind of "pulling" force. Ropes, strings, chains, rods, beams, and springs are all capable of exerting tension forces on other objects.

Like normal forces and static friction forces, tension forces present a dynamic response to a stimulus. If you pull on a rope, the rope pulls back on you. If you pull the rope harder, the rope pulls you harder. Because of this, there is no general formula for the strength of a tension force. However, there are some rules of thumb for dealing with ropes. Usually the ropes we encounter in physics have negligible mass compared to the objects they're

Don't be deceived by the apparent similarity between Equation 224 and Equation 223. These two equations have very different interpretations. Be able to explain the difference.

Another way to express Equation 224 is to say that static friction will break when the ratio f_s/n exceeds a certain value, namely μ_s.

FIGURE 239.
A bucket of water held just above the ground.

pulling on. When the mass of the rope is negligible, the rope can be considered to pull with the same force of tension on each end. For $\mathbf{F}_{\text{on one end}} - \mathbf{F}_{\text{on other end}} = \mathbf{F}_{\text{net on rope}} = m_{\text{rope}}\mathbf{a}_{\text{rope}} \approx 0$. Therefore, $\mathbf{F}_{\text{on one end}} = \mathbf{F}_{\text{on other end}}$.

> A massless rope/spring/rod/beam pulls with the same force of tension on each end.

Spring Forces

If you wanted to, you could use a shoelace to pull a sled. But you couldn't use a shoelace to *push* a sled! Shoelaces, ropes, chains, and strings can only pull. Springs are more versatile: They can push *or* pull. In this sense, springs are less like ropes and chains and more like rods and beams, because rods and beams can also both push and pull. However, springs also differ from rods and beams in that the length of a spring can change substantially when the spring is stretched or compressed.

An ideal spring has just two important characteristics: its rest length and its stiffness. The rest length, denoted ℓ_0, is the length of the spring when it's not being stretched or compressed—in other words, when nothing is pulling or pushing on the ends of the spring. Figure 240 shows a person lying on a bed, compressing the springs. For a spring, ℓ_0 refers to its resting length and ℓ refers to its length at other times (as here in the case of compression).

The stiffness of a spring is a measure of how difficult it is to stretch or compress the spring beyond its resting length. The stiffness of a spring is described in numerical terms by its "spring constant." The spring constant is usually denoted k and has units

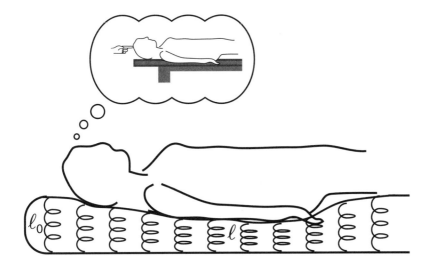

FIGURE 240.
A person lying on
a bed, compressing
the springs.

of newtons per meter. A large value of k (e.g., 1,000 N/m) indicates a very stiff spring, such as a shock absorber in a car. A small value of k (e.g., 1 N/m) indicates a very "mushy" spring, such as a Slinky toy.

When you pull on the end of a spring, you are of course exerting a force on the spring. But the spring is exerting an equal force on you. (If you doubt it, imagine attaching the end of the spring to your tongue post!) There is an easy formula for the strength of a spring force:

$$F_{\text{by spring}} = k|\ell - \ell_0| \, . \tag{225}$$

In this formula, k is the spring constant of the spring, ℓ_0 is the resting length of the spring, and ℓ is the actual length of the spring under the conditions being analyzed. If you are stretching or compressing the spring beyond its resting length, the actual length ℓ will differ from the rest length ℓ_0 by some amount, and, based on Equation 225, this difference will determine the strength of the resulting spring force.

Note that in Equation 225 $F_{\text{by spring}}$ is the *magnitude* of the spring force vector $\mathbf{F}_{\text{by spring}}$. Equation 225 doesn't tell you the *direction* of a spring force. But that's better said in words anyway: An ideal spring under compression pushes outward equally at both ends, and an ideal spring under tension pulls inward equally at both ends. This is just common sense, and together with our magnitude formula of Equation 225, it completes our basic theory of spring forces.

FIGURE 241.
The drag force exerted on a swimming tuna by the surrounding water.

FIGURE 242.
The drag forces exerted on a hiker by water moving past her feet.

Drag Forces

Drag forces arise when an object moves through a resistive fluid, such as air or water. Drag forces also arise when the target object is stationary and fluid moves past it—as happens in a wind tunnel.

Think of a tuna swimming in the ocean. As the tuna moves through the water, the water exerts a drag force on it, as shown in Figure 241. The drag force on a moving object always points directly against the velocity vector of the moving object.

When the fluid itself is moving and the target object is stationary, the drag force points with the velocity vector of the fluid. See Figure 242.

» WORKED EXAMPLE 3

Draw a force diagram for a skydiver falling to earth. Draw your diagram for a moment in time before the skydiver opens her parachute.

» SOLUTION 3

As usual, it is up to us to decide what forces act and what forces are important enough to include. Let's go through our list of kinds of force and see which ones we want to include.

- *Gravitational force.* Obviously the earth exerts a gravitational force on the skydiver. We'll include this in the diagram. (But we'll neglect any gravitational forces exerted by other celestial bodies!)

- *Electrostatic force.* Hmm. I don't see any charged objects in this problem. No electrostatic forces.

- *Magnetic force.* No magnets or current-carrying wires, either.

- *Normal force.* With any luck, a skydiver in free fall isn't bumping into any hard surfaces such as skyscrapers or tree branches. So we shouldn't include any normal forces in the diagram.

- *Friction force.* The skydiver isn't touching any surfaces (as noted earlier), so there aren't any friction forces acting on her.

- *Tension force.* The skydiver is not hanging from a chain, swinging on a rope, or anything like that. So there are no tension forces acting on her.

- *Spring force.* Unless we're talking about bungee jumping here, there are no springs attached to the skydiver. So don't draw any spring forces.

- *Drag force.* Drag forces from the air can be significant in this situation—especially as the skydiver attains a high speed. So we'll include the drag force in our analysis.

- *Lift.* If the skydiver were wearing a suit with winglike structures on the sleeves, there might very well be a lift force pulling upward on her. But let's keep things simple and assume that she doesn't have any fancy sleeves.

- *Thrust.* We'll assume that the skydiver isn't paddling with her arms or wearing a jetpack. So no thrust force is pushing the skydiver around.

Altogether, then, the forces worth including are shown in Figure 243. From the diagram, it looks as if the skydiver is speeding up.

◇

The Dependence of Drag Forces on Speed

If you've ever stuck your hand out the window of a moving car, you know that the drag force from the air is stronger when the car is moving faster. This suggests that the drag force depends in some

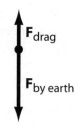

FIGURE 243. Solution to Worked Example 3. A force diagram for a skydiver falling downward.

way on the speed of the object moving through the fluid. But how, exactly?

This turns out to be an extremely complicated question, but we shall take a much-simplified view of things. On the simplest analysis possible, it turns out that there are actually two contributors to the total drag force, and each depends on the speed in its own way.

Viscous Drag

The first contribution to the drag force is what is known as *viscous drag*. Viscous drag is caused by the "stickiness," or *viscosity*, of the fluid you're moving through. Fluids like corn syrup are highly viscous; they flow slowly, and they present a great deal of viscous drag to any object moving through them. Maple syrup has a much lower viscosity than corn syrup, as you can tell by how quickly it runs over your pancakes.

Viscous drag forces are proportional to the speed of an object as it moves through the fluid:

$$F_{\text{viscous drag}} = bv.$$ (226)

in this formula, v is the speed of the object that is moving through the fluid, and b is a positive constant with units of kg/s.

Form Drag

The second contributor to the total drag force is what is known as *form drag*, or sometimes *pressure drag*. Form drag arises for the simple reason that in order to move through a fluid, you have to push some of the fluid ahead of you, like a snowplow.

Form drag is proportional to the *square* of the speed of the object as it moves through the fluid:

$$F_{\text{form drag}} = hv^2,$$ (227)

where v is the speed of the object that is moving through the fluid and h is a positive constant with units of kg/m.

Unlike viscous drag, form drag is heavily dependent on the "aerodynamic" profile or shape of the moving object. Its dependence on shape, or form, is what gives this force the name "form drag."

Drag Forces When the Fluid Is Moving, Too

In a wind tunnel, the fluid is moving and the target object remains still. In this case you can still use Equations 226 and 227 to analyze the drag forces at work. Just interpret the v as the speed of the fluid flow. More generally, if the fluid and the target object are *both* moving, the v in the formulas refers to the *relative speed* between the moving object and the fluid.

The Reynolds Number

In general, when an object moves through a fluid, viscous drag and form drag both act on the object. But in many cases, one kind of drag will be much stronger than the other, and we can safely neglect the smaller contribution entirely. The criterion for deciding which kind of drag force dominates is known as the *Reynolds number,* denoted Re.

The Reynolds number is a ratio that compares the importance of form drag (pushing a mass of fluid ahead of you) to the importance of viscous drag (dragging fluid with you):

$$\text{Re} \simeq \frac{\rho A v^2}{\mu L v} \simeq \frac{\rho d v}{\mu}, \tag{228}$$

where d is a rough length scale of the moving object, ρ is the mass density of the fluid, μ is the dynamic viscosity of the fluid, and v is the speed of the moving object. When the Reynolds number is low (Re < 100 or so), viscous drag dominates and the overall drag force is proportional to the speed. For higher Reynolds numbers, form drag dominates and the overall drag force is proportional to the square of the speed (albeit with a numerical drag coefficient C_D that depends strongly on the Reynolds number).

» WORKED EXAMPLE 4

A tuna swimming directly against a current flowing at 3 m/s manages to achieve a forward speed of 2 m/s relative to the sea floor. Estimate the rough order of magnitude of the drag force acting on the tuna.

In Equation 227, h is actually given by the product $\frac{1}{2} C_D \rho A$, where ρ is the mass density of the fluid (kg/m^3), A is the cross-sectional area of the object moving through the fluid, and C_D is the dimensionless "drag coefficient." The drag coefficient depends on the detailed shape of the object, as well as the speed v: $C_D = C_D(v)$.

»SOLUTION 4

First we'll estimate the Reynolds number. This will tell us whether viscous drag or form drag dominates.

To estimate the Reynolds number for a tuna swimming through seawater, we need the density of seawater ($\rho \simeq 1,000$ kg/m^3), the rough linear size of the tuna ($d \simeq 1$ m), the speed of the tuna relative to the water ($v = 2$ m/s $+ 3$ m/s $= 5$ m/s), and the viscosity of seawater ($\mu \simeq 10^{-3}$ at 20° C). These values give a Reynolds number of about

$$\text{Re} \simeq \frac{1,000 \cdot 1 \cdot 5}{10^{-3}} \tag{229}$$

$$\simeq 5 \times 10^6. \tag{230}$$

This is large enough that form drag dominates entirely.

Now to estimate the strength of the form drag force acting on the tuna using Equation 227. For this we need the cross-sectional area of the tuna ($A \simeq 0.1$ m^2) as well as the drag coefficient C_D appropriate to a fish-shaped object at a Reynolds number of 10^7 ($C_D \simeq 0.01$). Using these values, the drag force is given by

$$F_{\text{form drag}} \simeq \tfrac{1}{2} \cdot 0.01 \cdot 1,000 \cdot 0.1 \cdot 5^2 \tag{231}$$

$$\simeq 10 \text{ N}. \tag{232}$$

That's a pretty hydrodynamic animal! ◇

The hydrodynamics of swimming is a fascinating and active research area in marine biology. For example, it turns out that the skin of dolphins has morphological adaptations that effectively lower the Reynolds number of the flow, reducing the drag force on the animals and conserving precious resources of energy. See www.zoology.ubc.ca/courses/bio325/keyfact3.html.

Recently it has also been argued that dolphins take advantage of fluid-dynamical effects to keep mothers and calves pressed close together when swimming at high speeds. The effect is similar to the way a small car can sometimes be attracted to an 18-wheeler in the next lane due to the fast-moving slipstream between the two vehicles. In the case of the dolphins, this is in addition to a drafting effect that is said to provide up to 90% of the thrust required for calves to keep up with their mothers [8]. See http://jbiol.com/content/3/2/8.

Reader's Notes

Below is space for you to summarize key points from this chapter.

FOCUSED PROBLEMS :: CHAPTER 16

Now that we have so many technical tools for working with different forces, the number of problems we can solve has become vast. But we defer this avalanche until Chapter 17 so that we can take time to discuss problem-solving strategies.

1. A car is driving up a straight 45° incline at a constant speed. The wheels are not slipping against the ground. Draw a force diagram for the car. Neglect air resistance.

2. A car is driving *down* a straight 45° incline at a constant speed. The wheels are not slipping against the ground. Draw a force diagram for the car. Neglect air resistance.

3. How do your force diagrams for Focused Problems 1 and 2 compare?

4. Consider a car of mass m driving on a flat, straight road. The car is very gradually speeding up. Show that the car will begin to spin its wheels when it reaches a speed of $\sqrt{\mu_s mg/h}$, where μ_s is the coefficient of static friction and h is from Equation 227. (Note: This question is purely hypothetical; the slippage would not happen at any speed actually attainable by a car! But the scenario would apply pretty well to a hippopotamus running underwater.)

SOLUTIONS CAN BE FOUND ON PAGE 409

CHAPTER 17

Strategies for Applying Newton's Laws

In this chapter I'll suggest some organized strategies that can help you in applying Newton's Laws to understand physical situations and solve challenging physics problems. But before we get started, let's talk about attitude.

The Right Attitude

There are some easy physics problems out there. "The net force acting on a cat is 5 N, and the cat is accelerating at 2 m/s². What is the mass of the cat?" (Answer: 2.5 kg.) If all you ever do are simple problems like this, you can easily get the impression that $\mathbf{F}_{net} = m\mathbf{a}$ is a two-out-of-three equation: Give me two out of the three quantities, and I'll give you the third. But $\mathbf{F}_{net} = m\mathbf{a}$ is not a simple two-out-of-three equation. It's not like "distance = rate × time" or "tip = bill × 15%." The equation $\mathbf{F}_{net} = m\mathbf{a}$ is actually shorthand for an entire worldview: a grand vision of a mechanistic universe in which material particles are animated by their mutual interactions. $\mathbf{F}_{net} = m\mathbf{a}$ takes for granted the entire force concept, which took many pages to describe in Chapter 10. $\mathbf{F}_{net} = m\mathbf{a}$ demands that we know what acceleration means—a subtle concept that we spent three chapters on during the first half of this book. And finally, $\mathbf{F}_{net} = m\mathbf{a}$ relates two vectors that vary with time, with all the mathematical subtlety that that implies.

Please understand this: In physics problems of any significance, you are never "solving" $\mathbf{F}_{net} = m\mathbf{a}$. Rather, you are using $\mathbf{F}_{net} = m\mathbf{a}$ as a *framework* to help you *analyze a situation.*

I am talking about an important problem-solving attitude here. The grade-school attitude toward problem solving is "You give me the givens, and I'll give you the unknown." That won't work in physics. In physics, I'll give you a *situation,* and you'll *analyze it.* What we're looking for is a full understanding of the situation at hand. The specific quantities the problem does or doesn't ask for are almost beside the point.

A General Strategy for Solving Physics Problems

A general strategy can help you get a handhold on a complicated problem that would otherwise leave you not knowing where to start. Even students who soak up physics like sponges have a hard time solving substantial problems until they learn good strategies. In fact, sometimes those students suffer even more in my classes, because their ability to do easier problems by ad hoc methods causes them to put off learning any disciplined strategies.

Here, then, for your approval, is a general strategy for using $\mathbf{F}_{net} = m\mathbf{a}$ to solve physics problems. It's the strategy I give to my students when I teach this subject. It takes most of a year for people to become fully comfortable with it. Some never even bother. But even in the early weeks of your study of physics, bits and pieces of the strategy can usually get you unstuck when you need a push.

Without further ado, here's the strategy. You'll notice that there is no talk of "solving" anything until a tremendous amount of work has already been completed! After looking at the strategy, we'll see how it plays out in an example.

1. For each target:
   a. Sketch the target's acceleration vector at the instant of time under consideration.

      Do this by thinking about how the target is moving at this moment in time (speeding up/slowing/turning).

      Make an assumption if you have to.
   b. Draw a force diagram showing all forces acting on the target.

      Give an algebra name to the magnitude of each force (we usually use letters like n, T, F_1, F_2, f, etc.).

    c. Based on the vectors in your force diagram, sketch the net force vector.

        Does the net force vector you sketched match the acceleration vector you sketched? If $\mathbf{F}_{net}$ and $\mathbf{a}$ don't point the same way, something's wrong with your force diagram or your acceleration vector.

    d. Repeat this process for each of the targets in the problem.

        Don't move on until you're comfortable with your force diagrams! Everything comes from the diagrams. Garbage in, garbage out.

2. Identify all of the Third Law pairs.

    This will help make sure you haven't missed any forces.

    Use the same algebra name for Third Law pairs, because they are equal in magnitude.

3. For each target:

    a. Generally, choose one of your x- and y-directions to be positive along the acceleration vector, with the other direction perpendicular to that.

        This is not a hard-and-fast rule but it often works, especially in harder problems.

    b. Break up any "diagonal" forces.

        Erase the diagonal forces in your diagram and replace them with their equivalent x- and y-component forces.

    c. Write down Newton's Second Law, $\mathbf{F}_{net} = m\mathbf{a}$, in each direction. You might do it in table form, like this:

$\sum F_x$	=	ma_x

$\sum F_y$	=	ma_y

    d. Fill in the left-hand sides of the tables using the algebra names from your force diagram.

e. Fill in the right-hand sides of the tables by looking at the acceleration vector you drew.

f. Repeat for the rest of the targets in the problem.

Don't move on until you have fully analyzed all of the targets in both x- and y-directions.

4. Solve the stated problem.

You may need to invoke auxiliary facts such as $f_s^{\text{break}} = \mu_s n$, $F_{\text{grav}} = GMm/r^2$, $a_\perp = v^2/r$, etc.

Count up your unknowns and your equations, to make sure you have enough information.

See Figure 244 for a smaller-format version of this strategy that you can photocopy.

Let's see how this strategy plays out in solving a fairly simple problem.

» WORKED EXAMPLE 1

As I child, I used to love to ride an amusement park ride called the Roundup. The Roundup is a big cylinder with a radius of 3 meters. You step down into the cylinder and stand against the inner edge on a platform. The cylinder starts to turn faster and faster. When the cylinder reaches a high enough rate of spin, the platform is retracted and you "stick" to the walls, thanks to a large coefficient of static friction, $\mu_s = 1.5$ or so. How fast must the Roundup be spinning in order for you to stick to the wall? (Give your answer in revolutions per minute.)

» SOLUTION 1

First, attitude. Forget what the problem is asking you to do. The last words you read were "How fast must the Roundup be spinning," etc. But if you go into this with the attitude of trying to find out something to do with a rate of spin, you'll be lost. Instead, relax. Draw some cartoons. We're going to get to know this situation. And then you can bet that the answers we want will fall into our laps.

Figure 245 shows two views of a person riding on the Roundup after the platform has fallen away. The person is our target. Fortunately, there is only one target in this problem.

Step 1 of the strategy is a list of what should be done for each target in the problem. So let's do all those things for our person.

Of course, you can solve this problem any way you want. And you could very likely solve it *much* faster than we're about to solve it! I am going to go through all the detail you'll find in the following pages only so I can demonstrate to you some of the tricks of the trade.

1. For each target:

    a. Sketch the target's acceleration vector at the instant of time under consideration.

    Do this by thinking about how the target is moving at this moment in time (speeding up/slowing/turning). Make an assumption if you have to.

    b. Draw a force diagram showing all forces acting on the target.

    Give an algebra name to the magnitude of each force (we usually use letters like n, T, F_1, F_2, f, etc.).

    c. Based on the vectors in your force diagram, sketch the net force vector.

    Does the net force vector you sketched match the acceleration vector you sketched? If $\mathbf{F}_{net}$ and $\mathbf{a}$ don't point the same way, something's wrong with your force diagram or your acceleration vector.

    d. Repeat this process for each of the targets in the problem.

    Don't move on until you're comfortable with your force diagram! Everything comes from the diagram. Garbage in, garbage out.

2. Identify all of the Third Law pairs.

   This will help make sure you haven't missed any forces. Use the same algebra name for both members of Third Law pairs, because they are equal in magnitude.

3. For each target:

    a. Generally, choose one of your x- and y-directions to be positive along the acceleration vector, with the other direction perpendicular to that.

    This is not a hard-and-fast rule, but it often works, especially in harder problems.

    b. Break up any "diagonal" forces.

    Erase the diagonal forces in your diagram and replace them with their equivalent x- and y-component forces.

    c. Write down Newton's Second Law, $\mathbf{F}_{net} = m\mathbf{a}$, in each direction. You might do it in table form, like this:

$\sum F_x$	=	ma_x

$\sum F_y$	=	ma_y

    d. Fill in the left-hand sides of the tables using the algebra names from your force diagram.

    e. Fill in the right-hand sides of the tables by looking at the acceleration vector you drew.

    f. Repeat for the rest of the targets in the problem.

    Don't move on until you have fully analyzed all of the targets in both x- and y-directions.

4. Solve the stated problem.

   You may need to invoke auxiliary facts such as $f_s^{break} = \mu_s n$, $F_{grav} = GMm/r^2$, $a_\perp = v^2/r$, etc.
   Count up your unknowns and your equations, to make sure you have enough information.

FIGURE 244. A smaller-format version of my strategy for solving physics problems.

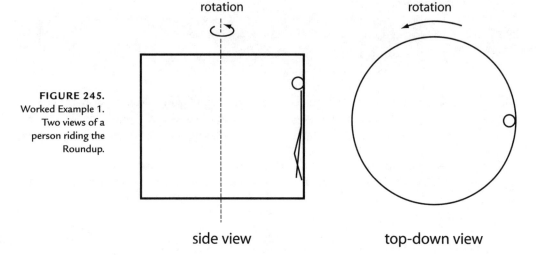

FIGURE 245.
Worked Example 1.
Two views of a
person riding the
Roundup.

Sketch the target's acceleration vector at the instant of time under consideration. We'll assume that the person is not speeding up or slowing down, just turning. (There *was* some speeding up until the platform fell away, but now we'll say that the ride is just maintaining its top speed.) Because the person is purely turning, the acceleration vector is always perpendicular to the velocity vector. I have sketched the acceleration vector in Figure 246. The velocity vector at the moment under consideration is also shown in the top-down view.

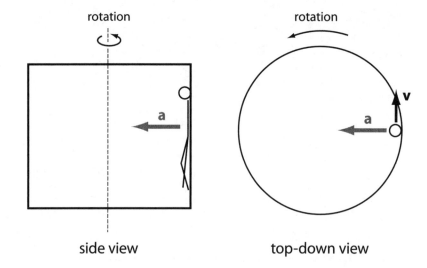

FIGURE 246.
Solving Worked
Example 1. The
acceleration vector
for the person riding
the Roundup.

Draw a force diagram showing all forces acting on the target. OK, what forces are acting on our person? For one thing, the earth pulls down on the person with a force that is gravitational in nature. And the person is touching the inner wall of the Roundup, so a normal force is probably also acting on the person. Normal forces always push away from the surface of the source object; that means that the normal force acting on the person points as shown in Figure 247. (We are just going to work with the side view from now on.)

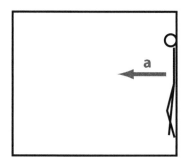
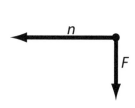

FIGURE 247.
Solving Worked Example 1. A proposed force diagram for the person riding the Roundup.

As suggested by the strategy, I'll give an algebra name to the magnitudes of both forces: F for the magnitude of the gravitational force and n for the magnitude of the normal force. Note that the variables F and n represent the *strengths* of the forces; they don't represent x- or y-components of the forces.

You're probably thinking that there are other forces acting on the person, but let's pretend we don't realize that and see what happens when we move on to the next step of the strategy.

Based on the vectors in your force diagram, sketch the net force vector. This means that we'll graphically add the force vectors in the diagram using the tip-to-tail method. When I do this, I get a result like that shown in Figure 248.

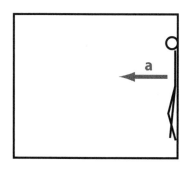
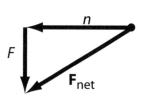

FIGURE 248.
Solving Worked Example 1. Adding the force vectors in the diagram using the tip-to-tail method.

Now we see a problem. $\mathbf{F}_{net}$ and $\mathbf{a}$ don't point the same way! This means that we're missing something. Either we're wrong about how this person is actually moving—that is, we've drawn the acceleration vector incorrectly—or else we're wrong about the forces that are acting on the person.

It's hard to imagine that we're wrong about the acceleration vector in this case; it's just simple circular motion, after all. So we're probably missing a force. In order for $\mathbf{F}_{net}$ to point horizontally (the same way as $\mathbf{a}$), we need an upward force to cancel the gravitational force. Indeed, there is an upward force acting here: static friction. If you picture yourself riding the Roundup, it's not hard to imagine that your shirt collar would creep up the back of your neck a little bit. This would be due to the force of static friction pulling your shirt upward, like a mother cat holding her kitten by the scruff of its neck. Figure 249 shows a new force diagram with the static friction force added. I chose the name f_s to represent the unknown strength of the static friction force.

FIGURE 249.
Solving Worked Example 1. A new force diagram with the static friction force added.

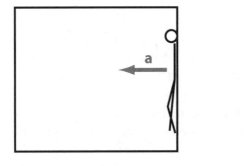
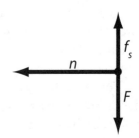

This is *static* friction, even though the person is moving with respect to the ground. The reason is that there's no *slippage* between the plastic of the Roundup and the fabric of the person's shirt.

This time, if we graphically add the vectors using tip-to-tail, it is plausible that the net force vector could look like the acceleration vector. Now $\mathbf{F}_{net}$ and $\mathbf{a}$ point the same way!

So now we're confident that we understand the forces at work. It's time to move on to step 2.

Identify all of the Third Law pairs. This step doesn't apply to the problem we're solving because we're analyzing only one target object. So we can move on to step 3.

Generally, choose one of your x- and y-directions to be positive along the acceleration vector, with the other direction being perpendicular to that. OK, following this rule of thumb, I set up my axes as shown in Figure 250.

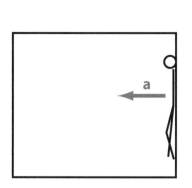

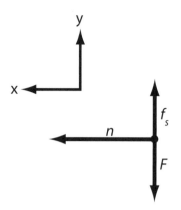

FIGURE 250.
Solving Worked Example 1. A coordinate system with one of the axes matching the direction of the acceleration vector of the target.

Break up any "diagonal" forces. All of the forces in the diagram point either purely along the *x*-axis, or purely along the *y*-axis. So there are no "diagonal" forces to break up.

Write down the Second Law $\mathbf{F}_{\text{net}} = m\mathbf{a}$ *in each direction.* Let's start with the *x*-direction:

$\sum F_x$	=	ma_x

It's your job to fill in both sides of the table. You have to fill in each side of the table *separately* before you can use the resulting equation.

To fill in the blank under $\sum F_x$, you look at the *x*-forces in the force diagram shown in Figure 250. I see only one *x*-force, namely the normal force *n*. Because *n* points *with* my *x*-arrowhead, *n* goes into the box with a plus sign:

$\sum F_x$	=	ma_x
$+n$		

Now for the box underneath ma_x. To fill in this box, you do something *totally different* than what you just did! You don't look at the forces on the diagram at all. Instead you look at the *acceleration vector*. The acceleration vector **a** points entirely in the positive *x*-direction. So the *x*-component of **a** is, well, some positive number, a_x. Caring not at all that we don't know what this number "a_x" is, we put it into the box:

$\sum F_x$	=	ma_x
$+n$		ma_x

Again, don't worry at this point about what's known and unknown. Remember: If you're worried about knowns and unknowns at this stage, you're simply in the wrong frame of mind.

Bringing down the equals sign, we have (ta-da!) our first equation:

(233)
$$n = ma_x.$$

It will often happen that you don't have a numerical value for an acceleration component a_x or a_y. So you will often just "bring down" an ma_x or ma_y when you are filling in the box on the right-hand side. But you should *never* bring down the symbols $\sum F_x$ or $\sum F_y$ when you are filling in the box on the left-hand side! Always fill in the box on the left with symbols from your force diagram.

Now for the *y*-direction. The Second Law for the *y*-direction is

$\sum F_y$	=	ma_y

To fill in the blank under $\sum F_y$, look at the *y*-forces in the force diagram in Figure 250. I see two *y*-forces, namely the static friction force f_s and the gravitational force F_{grav}. Because f_s points *with* my *y*-arrowhead, f_s goes into the box with a plus sign; since F_{grav} points *against* my *y*-arrowhead, F_{grav} goes into the box with a minus sign:

$\sum F_y$	$=$	ma_y
$+f_s - F_{grav}$		

Now for the box underneath ma_y. To fill in this box, once again look at the acceleration vector. The acceleration vector $\mathbf{a}$ points entirely in the positive x-direction. So the y-component of $\mathbf{a}$ is zero. This goes into the box:

$\sum F_y$	$=$	ma_y
$+f_s - F_{grav}$		0

Bringing down the equals sign, we have (oh joy!) our second equation:

$$f_s - F_{grav} = 0. \tag{234}$$

Solve the stated problem. We are finally at the solving stage. Our analysis of the situation has led to two equations:

$$n = ma_x \tag{235}$$
$$f_s - F_{grav} = 0. \tag{236}$$

Contrary to what you may have been taught, the first step in solving the problem is not to start playing with these equations. Instead, the first step is to identify which symbols you know the values of and which symbols you don't know the values of.

I count five different symbols: n, m, a_x, f_s, and F_{grav}. We don't know the values of any of them! Two equations in five unknowns cannot be solved. So there must be something else we can do.

Here is where those "auxiliary facts" come in. This is where we get to use our specialized knowledge about how different forces act. For example, the gravitational force on the person riding the Roundup has magnitude $F_{grav} = mg$. So we can put this into the system of equations:

$$n = ma_x \tag{237}$$
$$f_s - mg = 0. \tag{238}$$

This helps, because now we have only four unknowns. (The value of g is known.) But of course that's still two unknowns too many.

Here is where we can bring in another auxiliary fact: Our a_x is none other than $a_\perp$, the part of the acceleration vector perpendicular to the velocity vector. Therefore, a_x is equal to v^2/r. Maybe it will help to put that in:

$$n = m\frac{v^2}{R} \tag{239}$$

$$f_s - mg = 0. \tag{240}$$

I have replaced the generic symbol r in the slogan v^2/r with the given value of the Roundup's radius, R. This seems useful, although we still have four unknowns!

We also have an auxiliary fact about static friction: "$f_s^{\text{break}} = \mu_s n$." If this formula is actually relevant, it looks like it would help a lot. For one thing, the formula has a μ_s in it, and the problem quoted us a value of μ_s. I wouldn't put it past those pesky problem designers to trick us by providing irrelevant information, but it seems like we're on to something.

The formula $f_s^{\text{break}} = \mu_s n$ tells us the value of a static friction force just at the threshold of slipping. Actually, this is what we want, because the problem asks for the rotation rate necessary to make the person riding the Roundup stick to the wall. That means the lowest rotation rate that would make our person stick to the wall. The implication is that if the Roundup were spinning any slower than this critical value, the person would slip down the side.

So we feel pretty good about replacing f_s in our analysis with $\mu_s n$ to key on the slipping point. This leaves us with

$$n = m\frac{v^2}{R} \tag{241}$$

$$\mu_s n - mg = 0. \tag{242}$$

Let's count. There are six symbols here (n, m, v, R, μ_s, and g). We know the values of three of them (μ_s, R, and g). So now we're down to three unknowns, with two equations.

It looks like we're stuck. But here is where experience comes in. In problems that involve gravity, it often happens that the mass of the target object cancels out. To see if that will happen here, we can try some algebra. The first equation gives n in terms of other stuff, so let's use that in the second equation:

$$\mu_s \left(m\frac{v^2}{R} \right) - mg = 0. \tag{243}$$

After doing some algebra, we find

$$v^2 = \frac{gR}{\mu_s}. \tag{244}$$

The mass of the person has indeed dropped out! Apparently the Roundup works just as well for riders large and small—an important design consideration for an amusement park ride.

By taking the square root of both sides of our last equation, we have the minimum speed required in order for the person to stick to the wall:

$$v = \sqrt{(9.8 \text{ m/s}^2)(3 \text{ m})/1.5} = 4.4 \text{ m/s}. \tag{245}$$

The problem asks us for a minimum rotation rate in revolutions per minute, so we just have to figure out how long it takes to execute a full revolution at a speed of 4.4 m/s:

$$\text{time} = \frac{\text{distance}}{\text{rate}} = \frac{2\pi R}{v} = 4.26 \text{ s}. \tag{246}$$

One revolution in 4.26 s is the same as 0.235 revolutions per second, or 14 revolutions per minute. ◇

A Closer Look at the Roundup

In the spirit of analysis, let's forget about the specifics of the question that was originally asked and take a closer look at what's going on in the Roundup. Equations 239 and 240 are the basic equations that govern the person riding the Roundup:

$$n = m\frac{v^2}{R} \tag{247}$$

$$f_s - mg = 0. \tag{248}$$

According to these equations, the static friction force must always equal the person's weight, independent of the ride's rotation speed. Meanwhile, the normal force begins with a value of zero right when the Roundup first begins moving ($v \approx 0$), and then the normal force gradually becomes larger. It is interesting to look at the *ratio* of the static friction force to the normal force:

(249)
$$\frac{f_s}{n} = \frac{gR}{v^2}.$$

The speed of the person can be expressed in terms of the rotation frequency Ω as $v = \text{distance}/\text{time} = 2\pi R \times \left(\frac{1}{\text{time}}\right) = 2\pi R\Omega$. This allows us to express the ratio of static friction to normal force as

(250)
$$\frac{f_s}{n} = \frac{g}{4\pi^2 R\Omega^2} = \frac{0.083 \text{ s}^{-2}}{\Omega^2}.$$

It is interesting now to plot the ratio of static friction to normal force as a function of the Roundup's rotation frequency Ω. See Figure 251.

FIGURE 251.
A graph showing the ratio f_s/n as a function of the Roundup's rotation frequency.

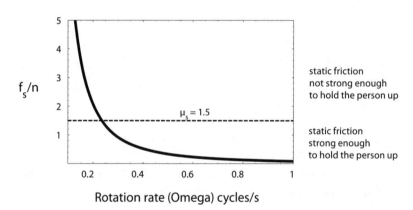

For low rotation frequencies—when the Roundup is just starting to "spin up"—the ratio f_s/n is greater than 1.5. This is a "forbidden zone" in which static equilibrium is impossible. The rider would slip without the platform under her feet. As the Roundup speeds up, the ratio f_s/n decreases until, at the critical value $\Omega = 0.235$ revolutions per second, the ratio drops below 1.5. From then on, as the frequency continues to rise, the ratio remains below the maximum allowed value of 1.5, and static friction is strong enough to hold the rider up. At the end of the ride, as Ω drops back below 0.235 revolutions per second, static friction is no longer strong enough to hold the rider up, and the platform must be restored.

Reader's Notes

Below is space for you to summarize key points from this chapter.

FOCUSED PROBLEMS :: CHAPTER 17

The Focused Problems for this chapter consist of a number of rigorous problems to give you practice in applying Newton's Laws.

1. Professor Snerd leaves the office at the end of the day with a book under his arm (mass of book $m = 0.2$ kg) and goes out to his car. He gets into the car and sets the book down flat on the seat next to him.

   a. Snerd takes off in a straight line and speeds up at a constant rate of $1\,\text{m/s}^2$. The book remains in place on the car seat. The coefficient of static friction between the book and the seat is $\mu_s = 0.8$. What is the force of friction on the book during the time when the car is speeding up?

      Note that although the book "remains in place" on the car seat, the book is actually accelerating (with respect to the ground). As discussed in Chapter 13, you should think about this problem not from Snerd's accelerating frame of reference but from the viewpoint of a spectator standing on the sidewalk watching the car go by.

   b. Snerd reaches a cruising speed of $v_0 = 15$ m/s, which he maintains over time, still continuing on a straight-line path. What is the force of friction on the book now?

2. A few blocks away from his office, Snerd turns a corner with radius of curvature r, maintaining his constant speed of 15 m/s the whole time. What is the tightest turn Snerd can take at this speed without the book slipping against the seat? (In other words, what is the smallest value of r for which the book will "remain in place" on the car seat? Take $\mu_s = 0.8$ once again.)

3. Little Billy is using a rope to pull a sled of mass $M = 3$ kg with a force of $T = 18$ N, applied at an angle of 30° above the horizontal. The sled moves horizontally across level ground. The

coefficient of kinetic friction is $\mu_k = 0.6$. Find the horizontal and vertical components a_x and a_y of the sled's acceleration vector.

4. Little Millie is using a rope to pull a sled of mass M along level ground. The coefficient of kinetic friction between the sled and the snow is μ_k. Millie pulls the rope at an angle α relative to the ground (Figure 252).

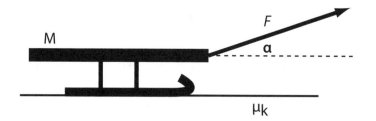

FIGURE 252.
Focused Problem 4.
Little Millie using a rope to pull a sled.

a. If Millie is to pull the sled at a constant speed along level ground, show that she must pull with a force

$$F(\alpha) = Mg \left(\frac{\mu_k}{\cos \alpha + \mu_k \sin \alpha} \right).$$

(251)

b. Choose reasonable numerical values for M and μ_k, and use a computer or graphing calculator to graph $F(\alpha)$ for values of α in the range $[0, 90°]$. You should find that there is an *optimum angle* α_{best} that minimizes the force required to drag something along the ground at a constant speed. Can you explain *why* there is an optimum angle? (Why is it unwise to pull at too steep an angle? Why is it unwise to pull at too shallow an angle?)

c. [Uses calculus.] If you know calculus, use it to show that $\alpha_{best} = \tan^{-1} \mu_k$.

5. A skier of mass $M = 100$ kg slides down a frictionless ski ramp. The ramp is in the shape of a circle with radius $R = 50$ m, as shown in Figure 253.

When the skier is at the point shown in the figure, her speed is 28.44 m/s. Knowing this, find all of the following quantities, in any order that makes sense to you:

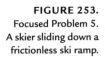

FIGURE 253.
Focused Problem 5.
A skier sliding down a
frictionless ski ramp.

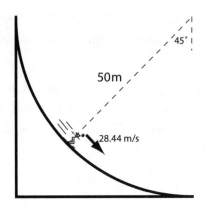

a. The radial component of the skier's acceleration vector, $a_\perp$.

b. The tangential component of the skier's acceleration vector, $a_\parallel$.

c. The magnitude of the force exerted on the skier by the ground.

6. Recall that your "weight" refers to the strength of the earth's gravitational pull on you, with magnitude 784 N near the earth's surface for an 80 kg person. Meanwhile, what a bathroom scale registers is how hard your feet are pushing against the scale.

a. Picture yourself standing on a scale at the South Pole. Sketch a force diagram for yourself. What does the scale read (in newtons) if your mass is 80 kg?

b. Now picture yourself standing on a scale at the Equator. Please recognize that when you are "standing still" at the Equator, you are actually accelerating. What does the scale at the Equator read, in newtons? Is this reading higher than your true weight or lower than your true weight?

c. Imagine what would happen if the earth were turning so fast that your feet barely grazed the scale. You would feel "weightless." This weightless feeling is exactly what astronauts are feeling when they orbit the earth so fast that their feet barely graze the floor of their spaceships. (It's actually not weightlessness, but rather "viselessness," as discussed in Chapter 16.) How short would the day have to be in order for the scale to read zero at the Equator?

7. Find the magnitude of the acceleration in the systems shown in Figure 254. These are standard block/pulley/inclined plane setups: massless frictionless pulleys and massless inextensible strings.

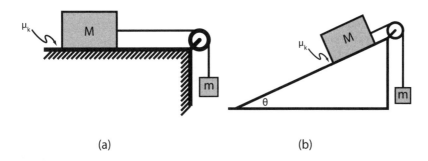

FIGURE 254.
Focused Problem 7.
Finding the magnitude of the acceleration in two systems.

(a)   (b)

You can give symbolic answers in terms of the symbols shown on the diagrams, or you can give numerical answers based on the following values: $M = 5\,\text{kg}$, $m = 1\,\text{kg}$, $\mu_k = 0.03$, $\theta = 30°$.

8. Figure 255 shows a standard block/pulley/inclined plane setup: massless frictionless pulley and massless inextensible string. The coefficient of static friction on the inclined plane is given as $\mu_s = 0.7$.

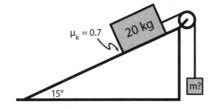

FIGURE 255.
Focused Problem 8.
Finding the value of the mass that will put the block on the verge of slipping.

Assume that the system is static (i.e., that neither block accelerates). Find the value of the mass m that will put the 20 kg block just on the verge of slipping. (Give a numerical answer.)

9. A car is moving on cruise control at a constant speed v_0. The road is extremely hilly, as shown in Figure 256.

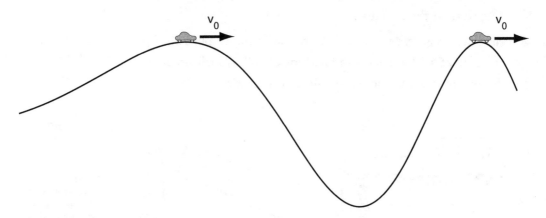

FIGURE 256.
Focused Problem 9.
A car moving at
constant speed
on a hilly road.

a. Draw two distinct and well-labeled force diagrams showing the forces acting on the car at the two instants of time depicted. In constructing your diagrams, consider all of the following kinds of forces and only these forces: gravity, normal force from the road, static friction from the road, and air resistance (viscous drag of the air on the car).

    (We shall assume that the car never spins its wheels; therefore, any friction force that may be present is necessarily static friction.)

b. Consider the very narrow hill (the second instant of time shown). Assume that at the top of the very narrow hill, the radius of curvature of the hill is given as R and the mass of the car is given as M. Find the speed at which the car just loses contact with the ground at the peak of the hill. You can give your answer in terms of any or all of the symbols M, g, and R, or you can give a numerical answer using $M = 1{,}000$ kg, $g = 9.8$ N/kg, and $R = 5$ m.

10. The equilibrium radius of a typical star is determined by competing forces: On the one hand, the self-gravitation of the star tends to compress the star to a small size; on the other hand,

this tendency to contract is balanced by the opposing tendency of a hot gas to expand.

As a very simplistic model of this situation, we may consider a spring with spring constant k and rest length L with equal masses M attached to either end, as shown in Figure 257.

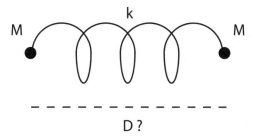

FIGURE 257.
Focused Problem 10.
A spring as a model of a star with an equilibrium radius determined by competing forces.

We think of the two masses as representing the left and right halves of the star. (So we are thinking of M as being on the order of 10^{30} kg.) These masses exert large gravitational forces on each other, tending to bring them together.

Meanwhile, we think of the spring as representing the tendency of the hot gas to push the star apart.

a. Find an equation that will determine the length D of the spring once the masses have come to rest. (Hint: Start by drawing a force diagram for one of the two masses. Another hint: People often want to say that the spring force has magnitude kD, but this is not right.)

b. Do not bother solving your equation, because it is cubic in the unknown D. However, *do* show that when M is too large, equilibrium is impossible: The star's self-gravity will cause it to collapse! This is a rough analogy to the way massive stars collapse and form black holes in the end stages of their evolution.

c. Show that when M is small enough, there are actually *two* equilibria for the system. This is a rough analogy to the two stable endpoints of stellar evolution for small and medium-sized stars: white dwarves and neutron stars.

11. A person is falling into a black hole of mass $M = 4 \times 10^{30}$ kg. At the moment of time we are considering, the person is at a distance $R = 10^6$ m from the black hole.

We will model the person as a pair of blocks of mass $m = 50$ kg each, connected by a massless rope of length 1m. In this model the rope is playing the role of the person's spinal column.

    a. Draw separate force diagrams for the two blocks that make up the person. Note that the gravitational forces on the two blocks will not be the same, because one block is closer to the black hole than the other block is.

    b. Find the tension in the rope.

    c. What unpleasant fate awaits someone who falls into a black hole?

SOLUTIONS CAN BE FOUND ON PAGE 411

Appendix: Derivation of Huygens's Formula

A Special Case: Uniform Circular Motion

Figure 258(a) shows two snapshots of an object moving at a constant speed v around a circular path of radius r. At time t_1, the velocity vector is $\mathbf{v}_1 = (+v, 0)$. At time $t_2 = t_1 + \Delta t$, the velocity vector is $\mathbf{v}_2 = (+v \cos \theta, +v \sin \theta)$. (We are taking x as positive to the right and y as positive upward.) Note that with a smaller angle θ, the vector difference $\mathbf{v}_2 - \mathbf{v}_1$ is more nearly vertical.

The acceleration vector $\mathbf{a} = \frac{d}{dt}\mathbf{v}$ is given approximately by

$$\mathbf{a} \approx \frac{\mathbf{v}_2 - \mathbf{v}_1}{t_2 - t_1} \tag{252}$$

$$= \frac{1}{\Delta t}[(v \cos \theta, v \sin \theta) - (v, 0)] \tag{253}$$

$$= \frac{1}{\Delta t}(v \cos \theta - v, v \sin \theta) \tag{254}$$

$$= \frac{v}{\Delta t}(\cos \theta - 1, \sin \theta). \tag{255}$$

During the time interval Δt between the two snapshots, the object will move a distance $d = v \Delta t$. The distance d is shown as a heavy dashed arc in Figure 258(a).

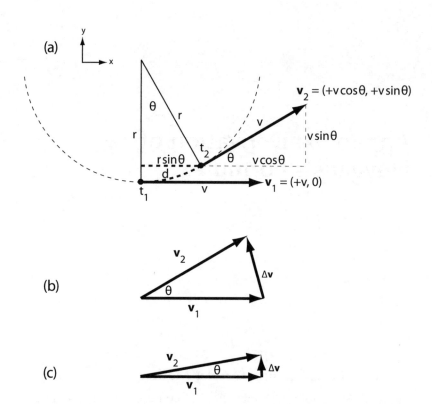

FIGURE 258.
Diagram for deriving
Huygens's Formula,
$a_\perp = \frac{v^2}{r}$, in the
special case of
uniform circular
motion. (a) The
velocity vectors at
two nearby times, t_1
and $t_2 = t_1 + \Delta t$.
(b) Subtracting the
two velocity vectors.
(c) What the velocity
vectors in (b) would
look like for a shorter
time interval Δt.

Rearranging this, we have $\Delta t = d/v$, so the ratio $\frac{v}{\Delta t}$ in Equation 255 is equal to $\frac{v}{d/v} = \frac{v^2}{d}$. Altogether, then, we have

$$\mathbf{a} \approx \frac{v^2}{d}(\cos\theta - 1, \sin\theta). \tag{256}$$

If the time interval Δt is brief, the angle θ will be small and the distance d traveled by the object will be approximately equal to $r\sin\theta$. See Figure 258(a). Putting $d \approx r\sin\theta$ into Equation 256 and simplifying, we have

$$\mathbf{a} \approx \frac{v^2}{r}\left(\frac{\cos\theta - 1}{\sin\theta}, +1\right). \tag{257}$$

When θ is small, the quantity $\frac{\cos\theta - 1}{\sin\theta}$ is negligibly small (see Table 6).

TABLE 6. Illustration of the Fact That as $\theta \to 0$, So Does $\frac{\cos\theta - 1}{\sin\theta} \to 0$

θ	$\frac{\cos\theta - 1}{\sin\theta}$
10°	−0.087
1°	−0.0087
0.1°	−0.00087
0.01°	−0.000087

In the ideal case of snapshots that are spaced infinitesimally close together, $\theta \to 0$, $\frac{\cos\theta - 1}{\sin\theta} \to 0$, and the exact value of the acceleration vector becomes

$$\mathbf{a} = \frac{v^2}{r}(0, +1) \tag{258}$$

$$\mathbf{a} = \left(0, \frac{v^2}{r}\right). \tag{259}$$

From Equation 259 we see that $a_\parallel = a_x = 0$ and that $a_\perp = a_y = \frac{v^2}{r}$. We have derived Huygens's Formula.

Generic Motions in the Plane

Next I'll show that Huygens's Formula works not only for uniform circular motion but indeed for just about any motion in the plane—whether the trajectory is circular or not and whether the speed is constant or not. (This derivation will use calculus.)

Consider an object moving in the x-y plane, with a position vector given by $\mathbf{r}(t) = (x(t), y(t))$. Figure 259 shows an example. Looking at the velocity vectors in the figure, we can imagine the object slingshotting around the curve, speeding up as it goes. The acceleration vector $\mathbf{a}$ is shown at the moment when the object passes through the origin. We shall show that the perpendicular component $a_\perp$ is given by $\frac{v^2}{r}$, where r is the radius of curvature at the moment in question.

For convenience, we have chosen our coordinate system so that the origin coincides with the position of the object at time $t = 0$ and so that the positive x-direction coincides with the direction of the object's velocity vector at time $t = 0$.

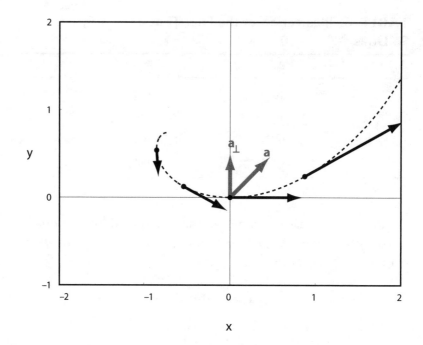

FIGURE 259.
An object moving on
a trajectory in the
x-y plane (dashed
curve). Four velocity
vectors are shown.
The acceleration
vector **a** is shown at
the moment when
the object passes
through the origin.

Over a short range of times near $t = 0$, Taylor's Theorem allows us to express $x(t)$ and $y(t)$ as polynomials:

$$(260) \qquad x(t) = vt + \frac{1}{2}a_x t^2 + \cdots$$

$$(261) \qquad y(t) = \frac{1}{2}a_y t^2 + \cdots.$$

The ellipses denote higher-order terms, which are negligible for small times. Figure 260 shows this Taylor approximation to the object's trajectory.

The constants in Equations 260 and 261 have been chosen so that the object is at the origin ($x = 0, y = 0$) at time $t = 0$, the velocity vector $\left(\frac{d}{dt}x, \frac{d}{dt}y\right)$ has components $(+v, 0)$ at time $t = 0$, and the acceleration vector $\left(\frac{d}{dt}v_x, \frac{d}{dt}v_y\right)$ has components (a_x, a_y) at time $t = 0$. We assume that $v > 0$; that is, we assume that at $t = 0$ the object is indeed moving. We shall also assume that $a_y \neq 0$; this will generically be the case for a motion in the plane.

The r in Huygens's Formula "v^2/r" is the radius of curvature of the trajectory at the origin. This is the radius of the circle that best approximates the trajectory at the origin. This "best-fit" circle will be a circle centered somewhere on the y-axis, passing through

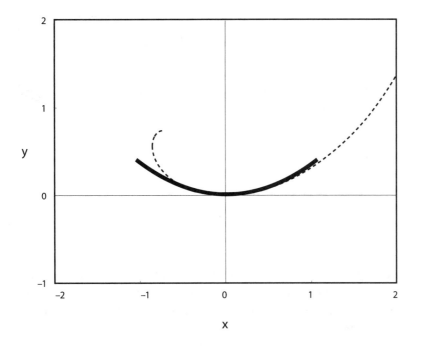

FIGURE 260.
The Taylor approximation to the trajectory of the object in Figure 259 near $t = 0$ (black curve).

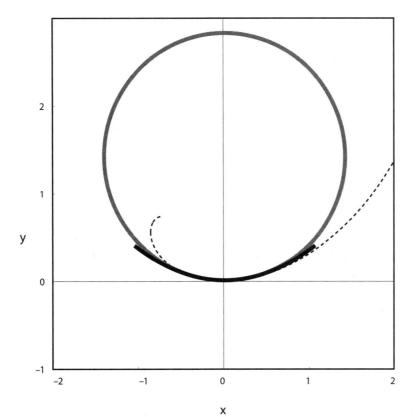

FIGURE 261.
The best-fit circle to the trajectory at $t = 0$.

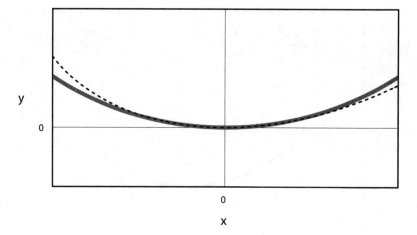

FIGURE 262.
Close-up of the origin showing the trajectory from Figure 261 (dashed curve) and the osculating circle (solid curve).

the origin and tangent to the x-axis. (See Figures 261 and 262.) The equation defining the best-fit circle will be $x^2 + (y - r)^2 = r^2$, where r is the radius of curvature. The radius of the circle shown in Figures 261 and 262 has been chosen to allow the circle to hug (or "kiss") the trajectory. The best-fit circle is known as the "osculating circle," from the Latin word *osculum*, meaning "kiss."

We can make the circle $x^2 + (y - r)^2 = r^2$ coincide with the trajectory as well as possible by inserting Equations 260 and 261 into the equation of the circle:

$$\left(vt + \frac{1}{2}a_x t^2 \right)^2 + \left(\frac{1}{2}a_y t^2 - r \right)^2 = r^2. \tag{262}$$

Expanding the squares and dropping negligibly small terms of order t^3 and t^4, we find

$$(v^2 - ra_y)t^2 = 0. \tag{263}$$

The equation holds at $t = 0$ because the circle intersects the trajectory at $t = 0$. The equation also automatically holds to the first order in time because the circle is tangent to the trajectory at the origin. However, to allow the circle to hug the trajectory to the second order in time, the term in parentheses must vanish:

$$v^2 - ra_y = 0. \tag{264}$$

Solving for a_y, which is also $a_\perp$ in this case, we find

$$a_\perp = \frac{v^2}{r}. \tag{265}$$

In this equation, v is the speed of the moving object at $t = 0$, and r is the radius of curvature of the trajectory at $t = 0$. We have not assumed that the speed v is holding constant, and we have not assumed that the trajectory is circular in shape.

If you worried that the best-fit circle might *not* have its center on the y-axis, insert Equations 260 and 261 into $(x - h)^2 + (y - k)^2 = r^2$, and discuss the implications. You should be able to deduce that $h = 0$, so our previous approach was actually fine.

Answers to Focused Problems

Chapter 1. Graphing Relationships

1. See Figure 263.

2. See Figure 264.

3. See Figure 265.

4. See Figure 266.

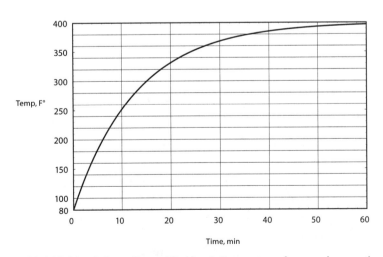

FIGURE 263. My solution to Focused Problem 1. Temperature of a casserole versus time.

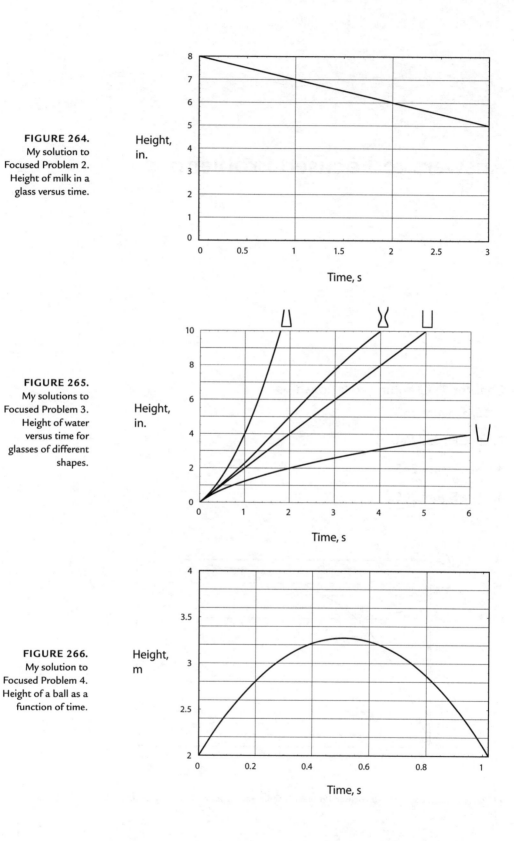

FIGURE 264.
My solution to
Focused Problem 2.
Height of milk in a
glass versus time.

FIGURE 265.
My solutions to
Focused Problem 3.
Height of water
versus time for
glasses of different
shapes.

FIGURE 266.
My solution to
Focused Problem 4.
Height of a ball as a
function of time.

Chapter 2. Rates of Change

1. This question was intentionally phrased in such a way that it has two reasonable answers. It's important to understand both of them. From one point of view, you might consider −9 bigger than −8. This is because −9 is larger in absolute value. For example, if I get punched in the mouth and lose 9 teeth, that's worse, or more severe, than if I were to lose only 8 teeth. Or let's look at another example. If the temperature outside is −9°, that's colder, or more severe, than if the temperature were only −8°.

   However, from another (and just as important) point of view, −8 is bigger than −9. This is because −8 lies to the right of −9 on the number line. So, whereas −9 is greater than −8 *in absolute value*, −8 is greater than −9 *as a real number*. And indeed, if the temperature were to change from −9° to −8°, we would certainly say that the temperature had increased, wouldn't we? This is a feature of negative numbers that's worth pointing out: *When negative numbers increase, their absolute value decreases.*

2. See Figure 267.

3. Answers are given in the figures that follow.
   a. See Figure 268.
   b. See Figure 269.
   c. See Figures 270–273.
   d. See Figure 274.

4. See Figures 268–274.

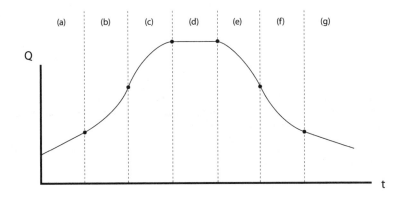

FIGURE 267. My answer to Focused Problem 2. Graph of a time-varying quantity.

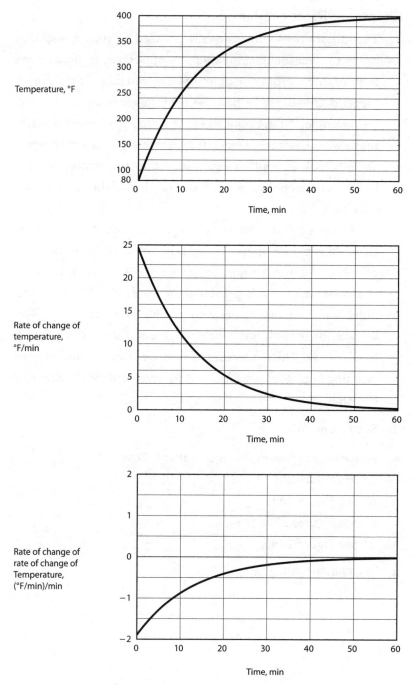

FIGURE 268. My answers to Focused Problems 3(a) and 4(a). Temperature of a casserole versus time, with the rate of change of the temperature versus time and the rate of change of the rate of change of the temperature versus time.

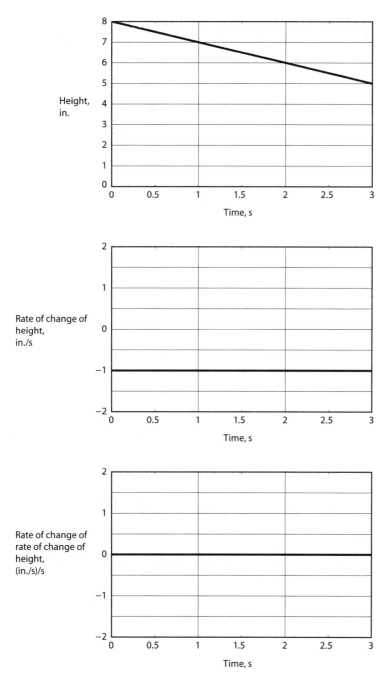

FIGURE 269. My answers to Focused Problems 3(b) and 4(b). Height of water in a glass versus time, with the rate of change of the height versus time and the rate of change of the rate of change of the height versus time.

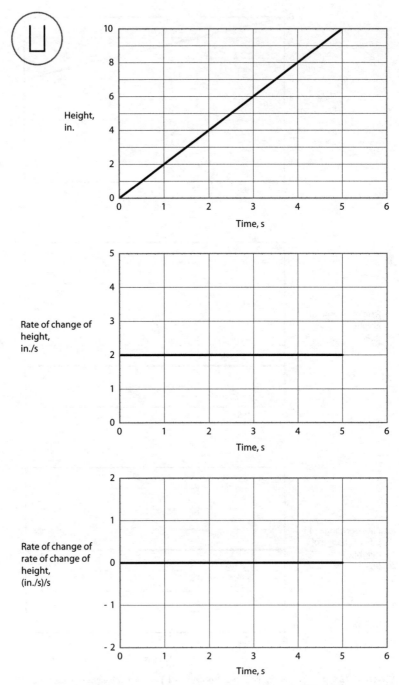

FIGURE 270. My answers to Focused Problems 3(c) and 4(c) for the first glass shown in Figure 26. Height of water in the glass versus time, with the rate of change of the height versus time and the rate of change of the rate of change of the height versus time.

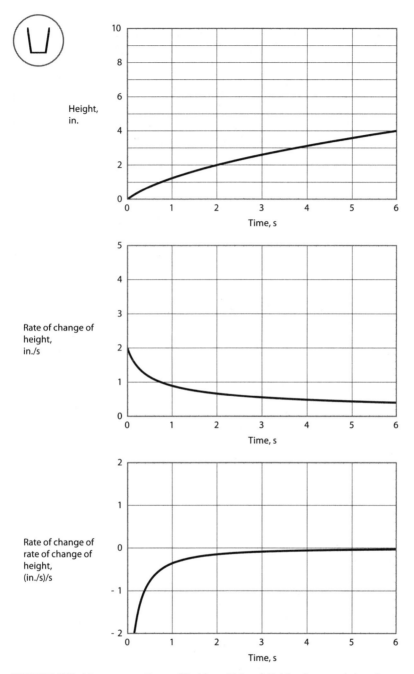

FIGURE 271. My answers to Focused Problems 3(c) and 4(c) for the second glass shown in Figure 26. Height of water in the glass versus time, with the rate of change of the height versus time and the rate of change of the rate of change of the height versus time.

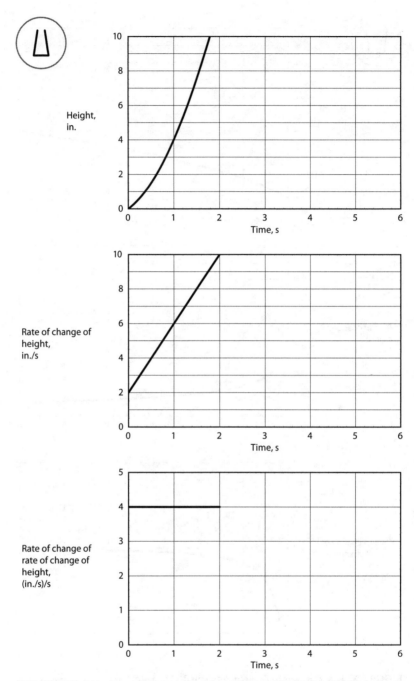

FIGURE 272. My answers to Focused Problems 3(c) and 4(c) for the third glass shown in Figure 26. Height of water in the glass versus time, with the rate of change of the height versus time and the rate of change of the rate of change of the height versus time.

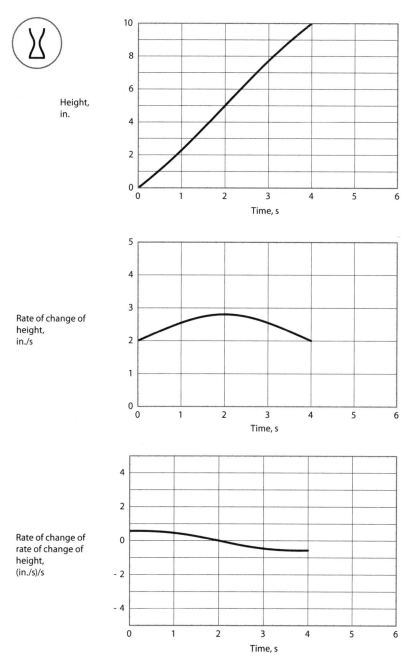

FIGURE 273. My answers to Focused Problems 3(c) and 4(c) for the fourth glass shown in Figure 26. Height of water in the glass versus time, with the rate of change of the height versus time and the rate of change of the rate of change of the height versus time.

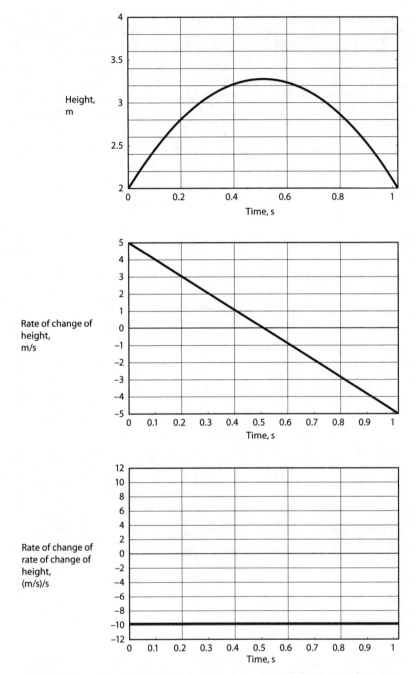

FIGURE 274. My answer to Focused Problem 3(d) and 4(d). Height of a ball versus time, with the rate of change of the height versus time and the rate of change of the rate of change of the height versus time.

5. I have drawn my estimated tangent line in Figure 275. Choosing two convenient points (40 wk, $9) and (44 wk, $8) on the tangent line, we calculate the rate of change as rise/run = ($9 − $8)/(40 wk − 44 wk) = $-\frac{1}{4}\frac{\$}{wk}$ = $-25\frac{cents}{wk}$. At time $t = 30$ wk, the stock price is falling at a rate of 25 cents per week.

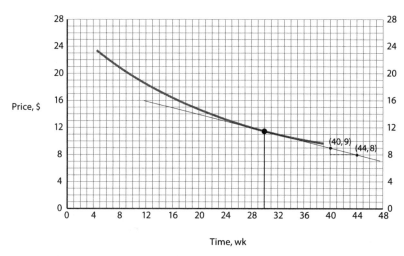

FIGURE 275. Focused Problem 5. Computing the rate of change of the stock price at time $t = 30$ wk.

6. $\frac{d}{dt}$[length of day] $\approx -0.049\frac{hr}{day} \approx -2.9\frac{min}{day}$.

7. Answers follow.
   a. $\Delta T = -6°$.
   b. $\Delta T = -8°$.
   c. $\Delta T = -6°$.
   d. $\Delta T = 0$. (At $t = 8$ hr, the temperature has returned to its original value.)
   e. $T_{min} = -22°$ at $t = 4$ hr.
   f. See Figure 276.

8. Answers follow.
   a. A careful estimate of $\frac{d}{dt}Q$ should be somewhere in the range from -1.5 to -2.5.

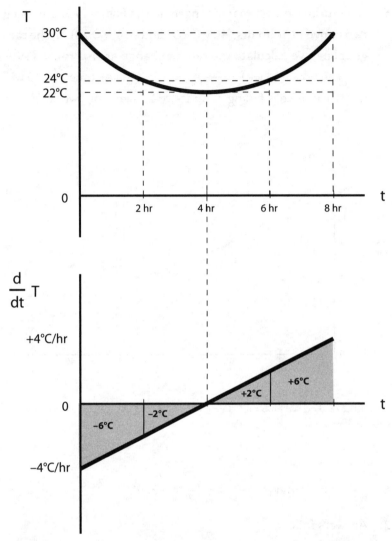

FIGURE 276. Focused Problem 7. Temperature as a function of time (top), determined from information in the rate-of-change graph (bottom) and assuming $T = 30°$ at time $t = 0$.

b. We can solve this part of the problem as follows:

$$\text{slope} = \frac{\text{rise}}{\text{run}} = \frac{\Delta Q}{\Delta t} \tag{266}$$

$$= \frac{Q(3 + \Delta t) - Q(t)}{(3 + \Delta t) - 3} \tag{267}$$

$$= \frac{10 + 4(3 + \Delta t) - (3 + \Delta t)^2 - (10 + 4(3) - (3)^2)}{\Delta t} \tag{268}$$

$$= \frac{10 + 12 + 4\Delta t - 9 - 6\Delta t - \Delta t^2 - 10 - 12 + 9}{\Delta t} \tag{269}$$

$$= \frac{4\Delta t - 6\Delta t - \Delta t^2}{\Delta t} \tag{270}$$

$$= \frac{-2\Delta t - \Delta t^2}{\Delta t} \tag{271}$$

$$= -2 - \Delta t. \tag{272}$$

c. We can solve this part of the problem as follows:

$$\text{slope} = \frac{\text{rise}}{\text{run}} = \frac{\Delta Q}{\Delta t} = \frac{Q(t + \Delta t) - Q(t)}{(t + \Delta t) - t} \tag{273}$$

$$= \frac{10 + 4(t + \Delta t) - (t + \Delta t)^2 - (10 + 4t - t^2)}{\Delta t} \tag{274}$$

$$= \frac{10 + 4t + 4\Delta t - t^2 - 2t\Delta t - \Delta t^2 - 10 - 4t + t^2}{\Delta t} \tag{275}$$

$$= \frac{4\Delta t - 2t\Delta t - \Delta t^2}{\Delta t} \tag{276}$$

$$= 4 - 2t - \Delta t \tag{277}$$

$$\rightarrow 4 - 2t. \tag{278}$$

d. Figure 277 shows a graph of $\frac{d}{dt}Q = 4 - 2t$ from $t = 0$ to $t = 6$. The rate of change starts out positive, drops to zero, then goes negative. The rate of change is greater, in absolute value, at the end (-8) than at the beginning $(+4)$.

These observations are consistent with the appearance of Figure 28. The curve starts off headed uphill, becomes shallower until it is momentarily horizontal, then becomes steeper and steeper downhill. The curve is steeper at the end than it was at the beginning.

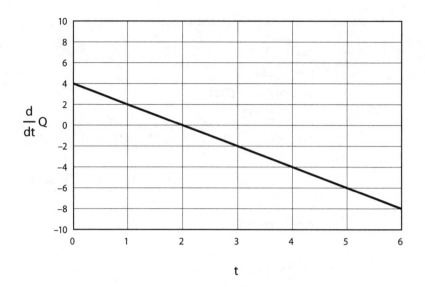

FIGURE 277.
Focused Problem 8.
A graph of the rate of
change $\frac{d}{dt}Q = 4 - 2t$
for a quantity varying
with time as
$Q(t) = 10 + 4t - t^2$.

Chapter 3. Introducing Position and Velocity

1. My scenario is that of a horse on a lead rope trotting around a circular path (Figure 278).

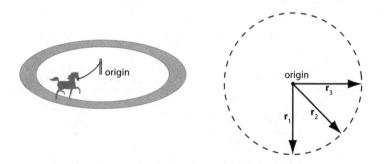

FIGURE 278.
Focused Problem 1.
The horse's distance
from the origin
remains constant,
but its position
vector changes over
time. The diagram on
the right is a
top-down view.

2. My scenario is once again that of a horse on a lead rope trotting around a circular path, this time with the further assumption that the horse trots at a constant speed (see Figure 279).

3. See Figure 280.

Chapter 4. Vectors

1. $F_{net,x} = +8.66$, $F_{net,y} = -1$. Magnitude: $F_{net} = 8.72$. Direction: 6.6° below the positive x-axis.

2. $F_{net,x} = -17.58$, $F_{net,y} = -14.75$. Magnitude: $F_{net} = 22.94$. Direction: 40° below the negative x-axis.

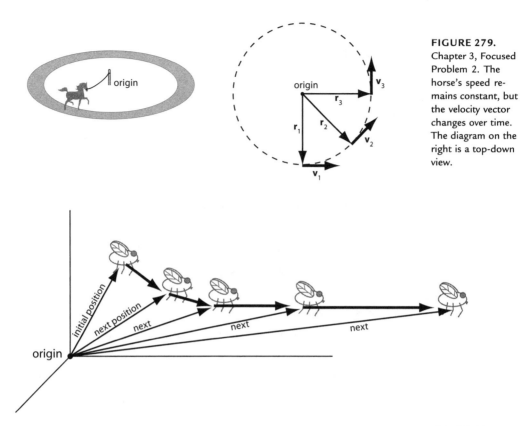

FIGURE 279.
Chapter 3, Focused Problem 2. The horse's speed remains constant, but the velocity vector changes over time. The diagram on the right is a top-down view.

FIGURE 280.
Chapter 3, Focused Problem 3. Sketches of the fly's velocity vectors at successive instants of time.

3. Magnitude: $F_{net} = 31.32$. Direction: $3.3°$ to the right of the positive y-axis. Because the vectors in this problem happen to be perpendicular to one another, I was able to determine the magnitude and direction of the vector sum without computing x- and y-components. Can you see how I did it? (Just in case you did compute the components, I'll quote them here so you can check your work: $F_{net,x} = +1.80$, $F_{net,y} = +31.27$.)

4. $F_1 = 7.07$.

Chapter 5. Position and Velocity, Revisited

1. 3.16 m.

2. $v \approx (-8 \text{ mph}, -4 \text{ mph})$; speed 8.94 mph, bearing $26.6°$ south of west.

3. Answers follow.

   a. 6 days.

b. 6 seconds.

c. 4 hours.

d. 9 seconds.

4. Answers follow.

a. From $\Delta x = v_x \Delta t$ we have $\Delta t = \Delta x / v_x$. With $\Delta x = 10$ ft and $v_x = (55 \text{ mph}) \sin 3°$, we find $\Delta t = 2.369$ s.

b. We have $\Delta y = v_y \Delta t = v_y (\Delta x / v_x) = \Delta x (v_y / v_x) = (10 \text{ ft}) \times (\cos 3° / \sin 3°) = 191$ ft. (Beware: some wrong approaches to this problem give answers that round off to 191 ft. To be sure you used a correct method, keep all of your calculator digits, and you should get 190.8113669. . . .)

5. Answers follow.

a. The lead boat will hang on for the win, with a margin of victory of only 0.13 s.

b. Let the trailing boat's bearing be denoted by an unknown angle θ north of west. (In part (a), θ was 22°.) Set the two boats' finish times equal:

$$\frac{100 \text{ m}}{10\frac{\text{m}}{\text{s}} \sin 20°} = \frac{110 \text{ m}}{10\frac{\text{m}}{\text{s}} \sin \theta} \tag{279}$$

$$\Rightarrow \quad \theta = \sin^{-1}(1.1 \sin 20°) \tag{280}$$

$$= 22.1°. \tag{281}$$

Chapter 6. Introducing Acceleration

1. The gas pedal is one of a car's accelerators—the one that speeds it up. In addition, the brake pedal is the accelerator that slows the car down, and the steering wheel is the accelerator that turns the car. (I suppose there's also the emergency brake.)

2. Answers follow.

a. A sprinter, just after the start of a 100-meter race (Figure 281(a)).

b. The same sprinter, just after crossing the finish line (Figure 281(b)).

c. A horse trotting around a lead circle at a constant speed (Figure 281(c)).

d. A roller coaster car gaining speed on the way down, just after cresting a hill (Figure 281(d)). Note the curvature of the track at the location of the car; this is where the turning component comes from.

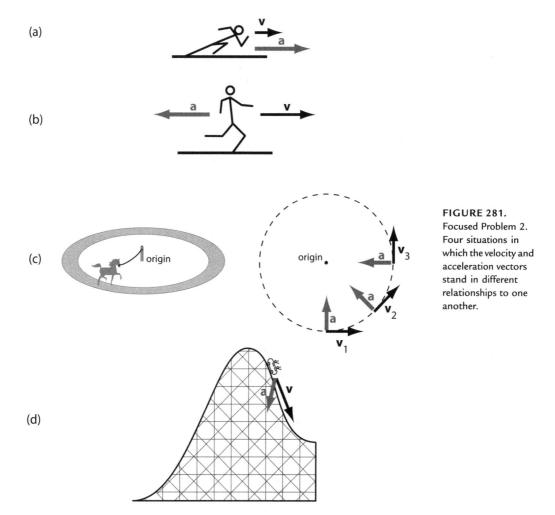

(a)

(b)

(c)

(d)

FIGURE 281. Focused Problem 2. Four situations in which the velocity and acceleration vectors stand in different relationships to one another.

3. See Figure 282.

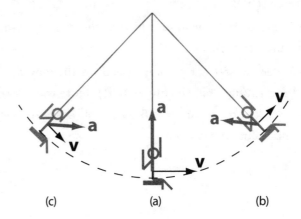

FIGURE 282.
Focused Problem 3.
Velocity and
acceleration vectors
for a tiny tot in a
swing.

(c)   (a)   (b)

4. See Figure 283. In my scenario, the penny falls straight down from rest and eventually attains a maximum speed because of air resistance. By the time the penny nears the ground, it is no longer speeding up, so its $a_\parallel$ is zero. And because the penny is moving in a straight line, its $a_\perp$ is zero throughout the motion. As the penny nears the ground, both components of its acceleration vector are zero.

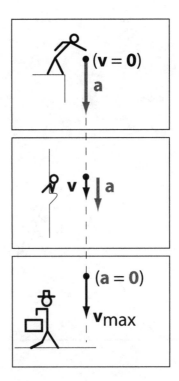

FIGURE 283.
Focused Problem 4.
Velocity and
acceleration vectors
for a penny dropped
from the Empire
State Building.

5. See Figure 284.

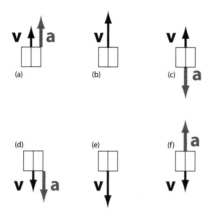

FIGURE 284. Focused Problem 5. Velocity and acceleration vectors for an elevator at different times during its motion. In diagrams (b) and (e), the acceleration vector is zero.

6. See Figures 285 and 286.

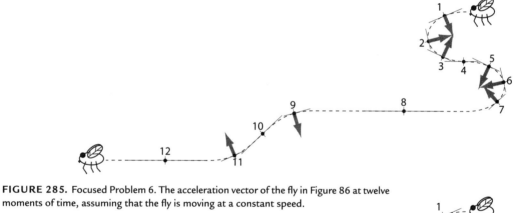

FIGURE 285. Focused Problem 6. The acceleration vector of the fly in Figure 86 at twelve moments of time, assuming that the fly is moving at a constant speed.

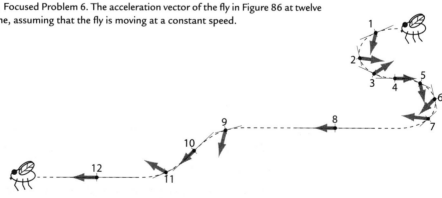

FIGURE 286. Focused Problem 6. The acceleration vector of the fly in Figure 86 at twelve moments of time, assuming that the fly is always speeding up. Notice the $a_\parallel$ at moments 4, 8, 10, and 12.

Chapter 7. Acceleration as a Rate of Change

1. Answers follow.

   a. See Figure 287.

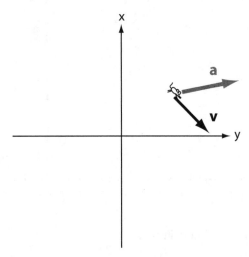

FIGURE 287. Focused Problem 1. Velocity and acceleration vectors for a mouse on the floor.

   b. Speed: $v = 4.2$ m/s. Direction of motion: southeast (45° south of east).

   c. Magnitude of acceleration: $a = 51$ m/s². Direction: 11° north of east.

   d. The angle between **v** and **a** is 45° + 11° = 56°. This means that the mouse is speeding up and turning.

   e. $v_{new,x} \approx (+3$ m/s$) + (+50$ m/s²$) \times (0.1$ s$) = +8$ m/s; $v_{new,y} \approx (-3$ m/s$) + (+10$ m/s²$) \times (0.1$ s$) = -2$ m/s; $v_{new} \approx 8.2$ m/s.

2. Answers follow.

   a. $\mathbf{v_{12}} \approx (-8$ mph, -4 mph$)$ at $t_{12} = 3{:}07{:}30$; $\mathbf{v_{23}} \approx (-16$ mph, -8 mph$)$ at $t_{23} = 3{:}22{:}30$; $\mathbf{a} \approx (-32\frac{mph}{hr}, -16\frac{mph}{hr})$. Magnitude of acceleration $a = 35.8\frac{mph}{hr}$, direction 26.6° south of west. See Figure 288.

   b. The acceleration vector points along both velocity vectors, which means that the ocean liner was speeding up and

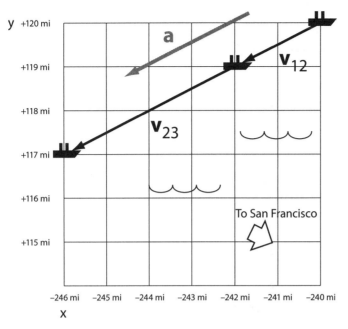

FIGURE 288. Chapter 7, Focused Problem 2. Computing the acceleration vector for a moving ship.

not turning. The speed increased from $v_{12} = 8.94$ mph to $v_{23} = 17.9$ mph.

3. See Figure 289. Note that the $v_x - t$ curve begins and ends with zero slope, because $a_x = 0$ at the beginning and end of the motion.

4. See Figure 290. Note that the $v_x - t$ curve begins and ends with zero slope, because $a_x = 0$ at the beginning and the end of the motion.

5. Answers follow.
   a. See Figure 291(a).
   b. See Figure 291(b). When the velocity is momentarily zero, the acceleration vector shows which way the velocity vector is *about* to point.

Chapter 8. Focus on a-Perp

1. See Figure 292.

2. See Figure 293.

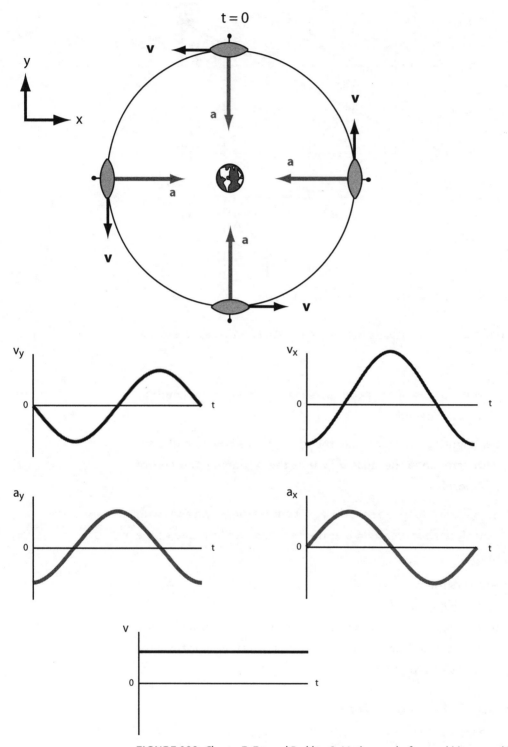

FIGURE 289. Chapter 7, Focused Problem 3. Motion graphs for an orbiting spaceship.

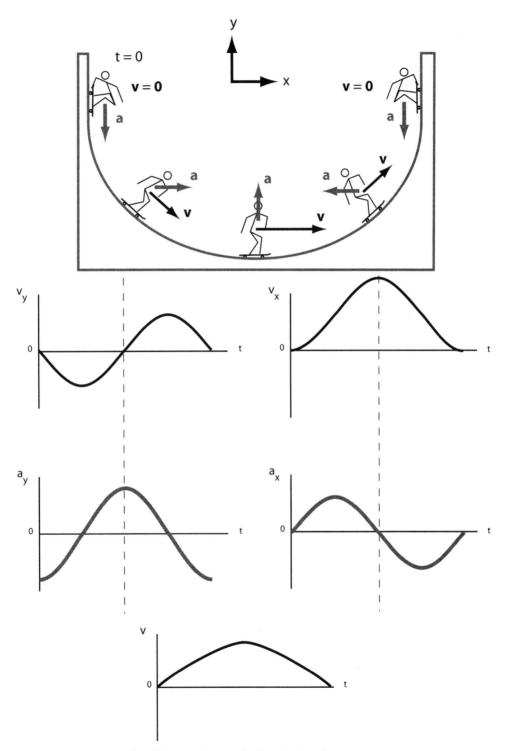

FIGURE 290. Chapter 7, Focused Problem 4. Motion graphs for a skateboarder.

FIGURE 291.
Chapter 7, Focused
Problem 5. Velocity
vectors at times t_0
and $t_0 + \Delta t$, with
the vector $\mathbf{a}(t_0)\Delta t$
representing their
difference. In case
(a), the velocity
vector is nonzero at
time t_0; in case (b),
the velocity vector is
zero at time t_0.

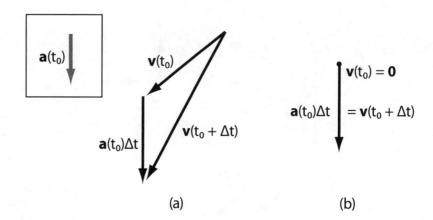

(a)   (b)

FIGURE 292.
Chapter 8, Focused
Problem 1. Velocity
and acceleration
vectors for a spiraling
penny.

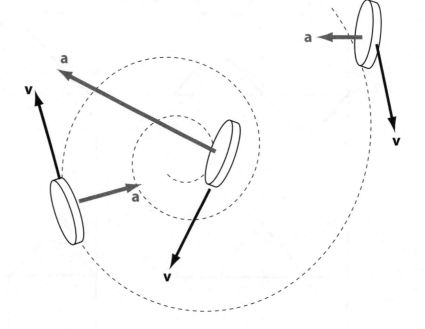

3. Answers follow.

    a. See Figure 294.

    b. $v = \frac{\text{distance}}{\text{time}} = \frac{2\pi r_0}{1 \text{yr}} = \frac{2\pi(1.5\times10^{11}\text{ m})}{365.25\times24\times60\times60\text{ s}} = 29{,}900 \text{ m/s}.$

    c. $a = \sqrt{a_\parallel^2 + a_\perp^2} = \sqrt{0 + a_\perp^2} = a_\perp = \frac{v^2}{r} = \frac{(29{,}900 \text{ m/s})^2}{1.5\times10^{11}\text{ m}} =$
    $0.006 \text{ m/s}^2.$

4. Answers follow.

    a. See Figure 295.

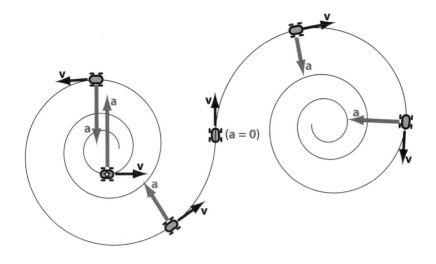

FIGURE 293.
Focused Problem 2. Velocity and acceleration vectors for a car driving on the indicated path.

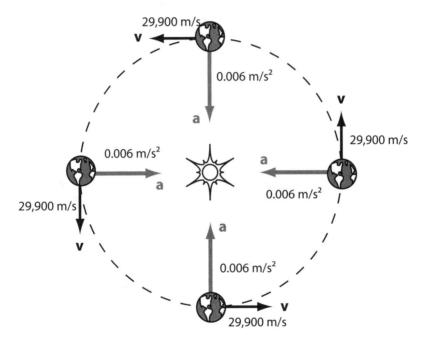

FIGURE 294.
Focused Problem 3. Velocity and acceleration vectors for the earth, labeled with numerical values. Distances are not to scale!

b. (i) $a_\perp = \frac{v^2}{r} = \frac{(5\text{ m/s})^2}{6\text{ m}} = 4.17\text{ m/s}^2$. (ii) $a_\perp = \frac{v^2}{r} = \frac{(8\text{ m/s})^2}{2\text{ m}} = 32\text{ m/s}^2$

c. (i) $a = \sqrt{(3\text{ m/s}^2)^2 + (4.17\text{ m/s}^2)^2} = 5.1\text{ m/s}^2$.
   (ii) $a = \sqrt{(3\text{ m/s}^2)^2 + (32\text{ m/s}^2)^2} = 32.1\text{ m/s}^2$.

5. See Figure 296.

6. See Figure 297.

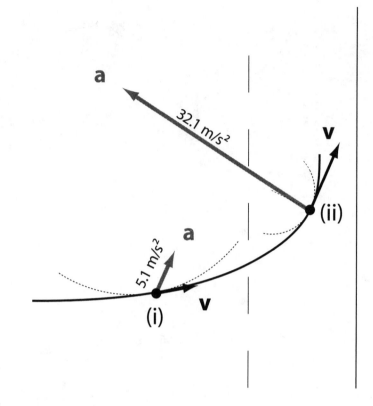

FIGURE 295.
Chapter 8, Focused
Problem 4. Accel-
eration vectors for a
car at two moments
of time.

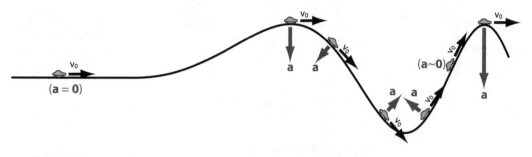

FIGURE 296.
Chapter 8, Focused
Problem 5. Velocity
and acceleration
vectors for a car
driving at constant
speed.

Chapter 9. Case Study: Straight-Line Motion

1. Answers follow.

   a. From just after $t = 2$ s to just before $t = 5$ s.

   b. From $t = 0$ to just before $t = 3.5$ s.

   c. From just after $t = 3.5$ s to $t = 7$ s.

   d. At $t = 3.5$ s.

   e. See Figure 298.

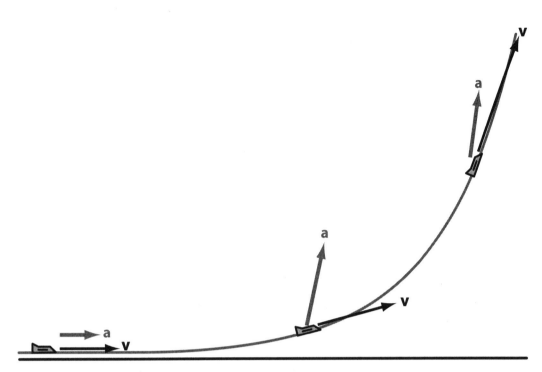

FIGURE 297.
Chapter 8, Focused
Problem 6. Velocity
and acceleration
vectors for a plane
taking off.

f. From $t = 0$ to just before $t = 2$ s and from just after $t = 3.5$ s to just before $t = 5$ s.

g. From just after $t = 2$ s to just before $t = 3.5$ s and from just after $t = 5$ s to $t = 7$ s.

h. At $t = 0$, Peanut is to the right of the origin, moving leftward, and speeding up. She passes through the origin at top speed and begins trying to reverse herself, coming to a momentary stop on the left side of the origin at $t = 3.5$ s, then speeding up again as she reverses course. Now heading rightward, she passes through the origin at top speed and begins slowing down. When the scene is over, she is still moving rightward, still slowing down, but reaching the end of the windowsill. (Looks like someone is going to fall off!)

2. $a_x = 22.3$ m/s^2; see Figure 299.

3. See Figure 300. My tangent line at $t = 3.5$ s passes through the grid points (3.3 s, 2.3 m) and (3.8 s, 1.4 m), so $v_x = \frac{d}{dt}x \approx$ (1.4 m $-$ 2.3 m)/(3.8 s $-$ 3.3 s) $= -1.8$ m/s. At $t = 0$, the

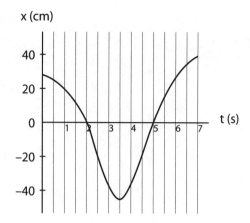

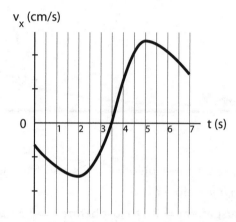

FIGURE 298. Focused Problem 1(e). $x - t$ and $v_x - t$ graphs.

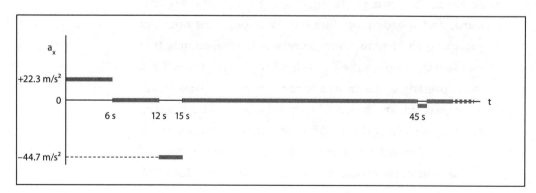

FIGURE 299. Focused Problem 2. $a_x - t$ graph for a dragster.

electron is 4 m away from the origin. The electron is at rest, but in the process of speeding up toward the origin. The electron continues toward the origin, moving faster and faster, ending up with a nearly infinite speed. The electron never slows down, stops, or switches directions.

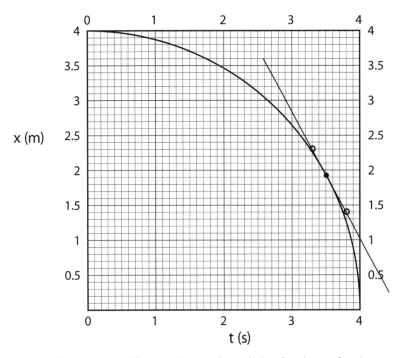

FIGURE 300. Focused Problem 3. Estimating the x-velocity of an electron from its $x - t$ graph.

4. See Figure 301. When the scene begins, the car is at the origin standing still. The car remains at rest until around $t = 1.5$ s, when it begins to speed up, heading in the positive x-direction. At around $t = 2$ s, the car reaches its peak speed and then begins slowing down. When its speed drops to about $1\,\text{m/s}$, the car holds that speed for the remainder of the scene. The x-acceleration of the car is highly variable, not constant at all.

5. See Figure 302. At $t = 0$ the taxicab is at rest but in the process of speeding up in the positive x-direction. The cab gains speed until its peak speed is reached at $t = 2$ s, at which point the cab starts to slow down, coming to rest at time $t = 4$ s. In order to achieve this kind of motion, the cabdriver would have to

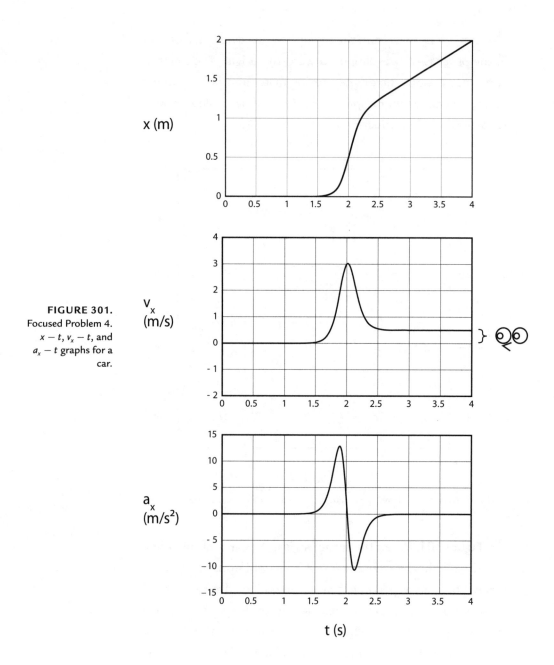

FIGURE 301.
Focused Problem 4.
$x - t$, $v_x - t$, and
$a_x - t$ graphs for a
car.

suddenly mash the gas pedal to the floor at time $t = 0$, then gradually bring the gas pedal back up, finally releasing it entirely at time $t = 2$ s and immediately moving his foot to the brake pedal; from $t = 2$ s to $t = 4$ s, he gradually presses the brake harder and harder, until at time $t = 4$ s the brake is mashed to the floor. (Having lived in New York City for a while, I can attest

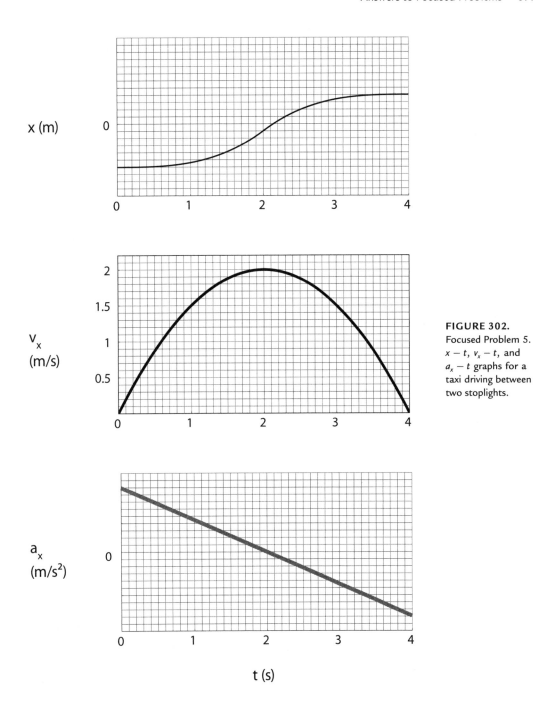

FIGURE 302.
Focused Problem 5.
$x - t$, $v_x - t$, and
$a_x - t$ graphs for a
taxi driving between
two stoplights.

that this is actually the way some NYC cabdrivers operate the pedals when going from one stoplight to the next: mash the gas, ease off, apply the brake, then mash the brake just as they come to a full stop.)

6. Answers follow.

   a. $v_x(0) = 10^5$ m/s. $v_x(5 \times 10^{-7}$ s$) = 8.66 \times 10^4$ m/s.
     $v_x(10^{-6}$ s$) = 0$.

   b. See Figure 303.

FIGURE 303.
Focused Problem 6.
$v_x - t$ and $a_x - t$
graphs for an
electron in a plasma.

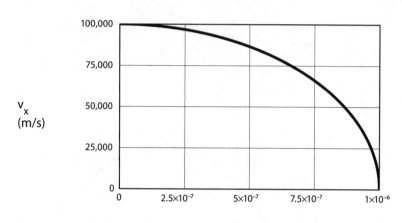

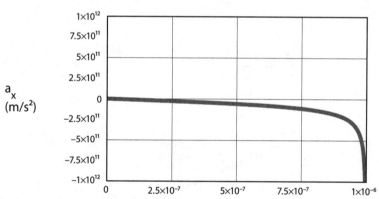

   c. See Figure 303.

   d. At time $t = 0$, the proton is moving in the positive x-direction with a speed of 10^5 m/s. The electron slows more and more suddenly, coming to rest at time $t = 10^{-6}$ s.

7. $\Delta x = 7.9$ cm.

8. Answers follow.

   a. See Figure 304.

   b. See Figure 305.

   c. They meet 59 m above the ground at $t = 6.0665$ s, so Neo arrives in time.

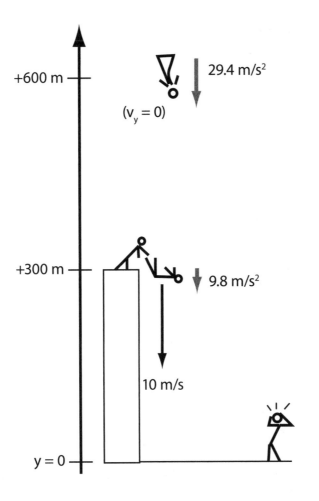

+600 m

29.4 m/s²

$(v_y = 0)$

+300 m

9.8 m/s²

10 m/s

y = 0

FIGURE 304.
Focused Problem 8.
Diagram of the
situation at $t = 0$
showing velocity and
acceleration vectors.

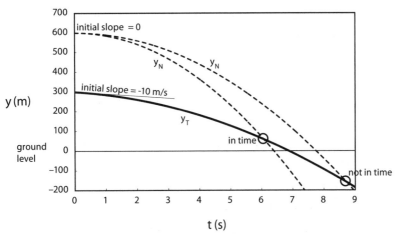

700

600   initial slope = 0

500

400   y_N y_N

300   initial slope = -10 m/s

y (m)   200

100   y_T

ground   0   in time
level
        -100

        -200   not in time

        0   1   2   3   4   5   6   7   8   9

t (s)

FIGURE 305.
Focused Problem 8.
Some $x - t$ graphs
for Trinity (solid
curve) and Neo
(dashed curves). Two
possibilities for Neo
are shown.

9. Answers follow.

a. See Figure 306.

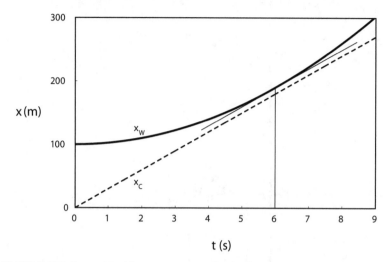

FIGURE 306. Focused Problem 9. $x - t$ curves for the cheetah and the wildebeest.

b. $t = 6$ s.

c. $x_W(t = 6\,\text{s}) - x_C(t = 6\,\text{s}) = 190\,\text{m} - 180\,\text{m} = 10\,\text{m}$. This is the closest the cheetah will get to the wildebeest. Can you see why?

d. On the $x - t$ curve, the time when the cheetah and the wildebeest have the same speed will be visible as the time when the *slopes* of the two position curves are the same. (See the thin tangent line in Figure 306.)

e. The easiest way to answer this question is to realize that if the cheetah had merely started out 10 m closer to the wildebeest, she would have caught her prey. Therefore, the critical head start is $D_{crit} = 100\,\text{m} - 10\,\text{m} = 90\,\text{m}$.

An alternative approach would be to reason as follows. The critical head start D_{crit} has the property that the cheetah catches the wildebeest just as the two animals reach the same speed. Regardless of the amount of head start, the two animals will reach the same speed at $t = 6$ s. So we can find D_{crit} by requiring that the cheetah catch the wildebeest at $t = 6$ s:

$$x_C(t = 6\text{ s}) = x_W(t = 6\text{ s}) \tag{282}$$

$$(30\text{ m/s})(6\text{ s}) = D_{\text{crit}} + \frac{1}{2}(+5\text{ m/s}^2)(6\text{ s})^2 \tag{283}$$

$$\Rightarrow \quad D_{\text{crit}} = 90\text{ m}. \tag{284}$$

A mathematically interesting way to reach the same conclusion is as follows. If we set $x_C = x_W$ using the given information, we get the equation $30t = 100 + 2.5t^2$. In standard form, this is $2.5t^2 - 30t + 100 = 0$. The quadratic formula gives $t = (30 \pm \sqrt{900 - 4 \times 2.5 \times 100})/5$. But the quantity under the square root is negative, so there are no solutions. (This was to be expected, because the two curves in Figure 306 do not actually intersect!)

The trouble under the square root sign comes from the large head start (100). If we decrease the head start, the quantity under the square root will become positive and the curves will intersect. The critical head start is the value D_{crit} for which the quantity under the square root is zero: $900 - 4(2.5)D_{\text{crit}} = 0$. This gives $D_{\text{crit}} = 90$ m as before.

10. Using $v_{f,x} = v_{i,x} + a_x t$, we have

$$v_{f,x}^2 - v_{i,x}^2 = (v_{i,x} + a_x t)^2 - v_{i,x}^2 \tag{285}$$

$$= v_{i,x}^2 + 2v_{i,x}a_x t + a_x^2 t^2 - v_{i,x}^2 \tag{286}$$

$$= 2a_x\left(v_{i,x}t + \frac{1}{2}a_x t^2\right) \tag{287}$$

$$= 2a_x(x_f - x_i) \tag{288}$$

$$v_{f,x}^2 - v_{i,x}^2 = 2a_x \Delta x. \tag{289}$$

In going from the third line to the fourth, I used Equation 103 in the form $x_f = x_i + v_{i,x}t + \frac{1}{2}a_x t^2$.

11. Answers follow.

a. Figure 307 shows $x - t$, $v_x - t$, and $a_x - t$ graphs for the plane, taking the origin of the coordinates to be at the starting point and taking the positive x-direction to be the forward direction along the runway.

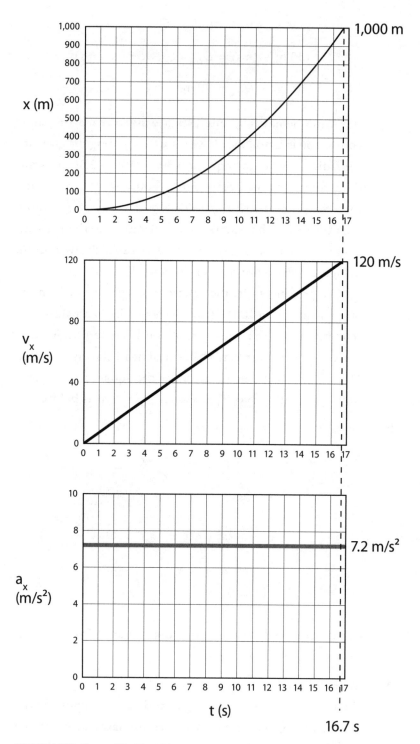

FIGURE 307. Focused Problem 11. $x - t$, $v_x - t$, and $a_x - t$ graphs for a plane reaching takeoff speed.

b. Equation 143 gives $(120 \text{ m/s})^2 - 0^2 = 2a_x(+1{,}000 \text{ m} - 0 \text{ m})$, and we solve for $a_x = +7.2 \text{ m/s}^2$.

c. Figure 307 is labeled with quantitative information. Table 7 shows the plane's x-velocity at various times, using $v_x(t) = 0 + (+7.2 \text{ m/s}^2)t$.

TABLE 7. The x-Velocity of the Plane at One-Second Intervals

t	$v_x(t)$
0	0
1 s	7.2 m/s
2 s	14.4 m/s
3 s	21.6 m/s
4 s	28.8 m/s
5 s	36.0 m/s
6 s	43.2 m/s
7 s	50.4 m/s
8 s	57.6 m/s
9 s	64.8 m/s
10 s	72.0 m/s
11 s	79.2 m/s
12 s	86.4 m/s
13 s	93.6 m/s
14 s	100.8 m/s
15 s	108.0 m/s
16 s	115.2 m/s
17 s	122.4 m/s

12. Answers follow.

a. See Figure 308. The graphs are drawn for the case in which Bob wins. My coordinate system has the origin $x = 0$ at the start of the race and $x = +10$ m at the end of the race.

b. Alice will finish the race when $x_A = 10$ m, that is, when $2 \text{ m} + (1 \text{ m/s})t = 10$ m, that is, at $t = 8$ s. By this time, Bob would

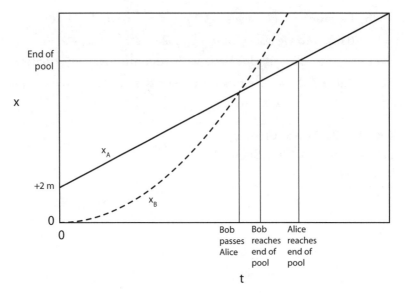

FIGURE 308.
Focused Problem 12.
Alice and Bob's first
swimming contest.

be at position $x_B(8 \text{ s}) = \frac{1}{2}(+0.5 \text{ m/s}^2)(8 \text{ s})^2 = +16 \text{ m}$, well
past the far end of the pool. So Bob will win. The sketch in
Figure 308 is qualitatively correct.

c. See Figure 309.

d. Starting from the equation $v_{x,B} = 0 + (0.5 \text{ m/s}^2)t$, write in
1.5 m/s for $v_{x,B}$ to find that Bob reaches his maxi-
mum speed at $t = 3$ s. By this time, he is at position
$x_B = \frac{1}{2}(0.5 \text{ m/s}^2)(3 \text{ s})^2 = 2.25 \text{ m}$. Meanwhile, Alice is at
position $x_A = 2 \text{ m} + (1 \text{ m/s})(3 \text{ s}) = 5 \text{ m}$.

e. Restart the clock when Bob reaches peak speed. Now $x_B = 2.25 \text{ m} + (1.5 \text{ m/s})t$ and $x_A = 5 \text{ m} + (1 \text{ m/s})t$. Alice reaches
the far wall when $x_A = 10$ m, that is, at $t = 5$ s. At this time,
Bob is located at $x_B = 2.25 \text{ m} + (1.5 \text{ m/s})(5 \text{ s}) = 9.75 \text{ m}$.
Alice wins by only 25 centimeters!

13. Answers follow.

a. See Figure 310. The change in v_x is given by the area under
the $a_x - t$ curve, $\Delta v_x = \text{base} \times \text{height} = T \times a_x = a_x T$.

b. See Figure 310. The change in x is given by the area under
the $v_x - t$ curve. This area is $\Delta x = (\text{area of rectangle}) + (\text{area of triangle}) = T \times v_{i,x} + \frac{1}{2}(T)(a_x T)$. This simplifies to
give Equation 144.

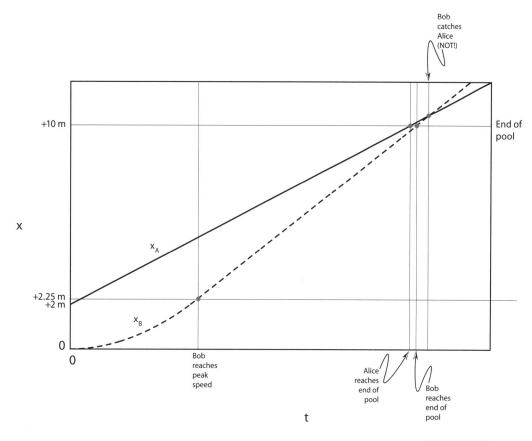

FIGURE 309.
Focused Problem 12.
Alice and Bob's
second swimming
contest.

14. One form of the Fundamental Theorem of Calculus states that for a differentiable function $f(t)$, $\int_{t_i}^{t_f} \frac{df}{dt}\, dt = f(t_f) - f(t_i)$. If we take $f(t)$ to be $x(t)$, $\frac{df}{dt}$ is $\frac{dx}{dt}$, that is, v_x, so the Fundamental Theorem gives $\int_{t_i}^{t_f} v_x\, dt = x(t_f) - x(t_i)$, that is, $\Delta x = \int_{t_i}^{t_f} v_x\, dt$. In other words, the change in position is the net signed area between the $v_x - t$ curve and the t-axis between t_i and t_f (this is the meaning of the symbol $\int_{t_i}^{t_f} v_x\, dt$).

15. Answers follow.

   a. The cab starts from rest because $v_x(0) = \frac{8v_0^2}{3D}\left(0 - \frac{2v_0}{3D}0^2\right) = 0$. The cab reaches maximum speed when $\frac{dv_x}{dt} = 0$, that is, when $1 - \frac{4v_0}{3D}t = 0$, that is, when $t = \frac{3D}{4v_0}$. The cab comes to rest again at $t_{\text{stop}} = \frac{3D}{2v_0}$, because $v_x\left(\frac{3D}{2v_0}\right) = \frac{8v_0^2}{3D}\left(\frac{3D}{2v_0} - \frac{2v_0}{3D}\left(\frac{3D}{2v_0}\right)^2\right) = 0$.

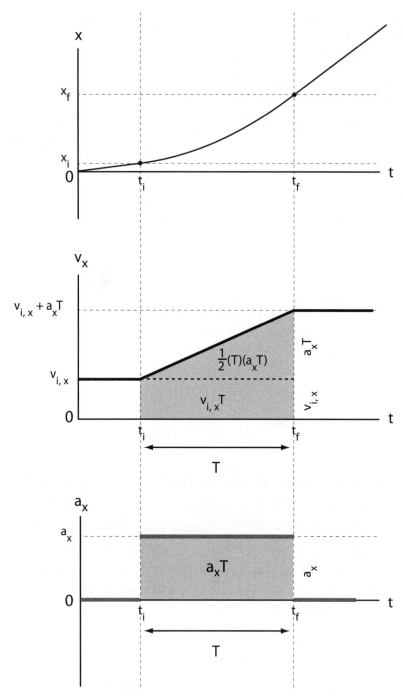

FIGURE 310. Focused Problem 13. Deriving Equation 144.

b. We can solve this part of the problem as follows:

$$\Delta x = \int_0^{t_{stop}} v_x(t)\,dt \tag{290}$$

$$= \int_0^{3D/2v_0} \frac{8v_0^2}{3D}\left(t - \frac{2v_0}{3D}t^2\right)\,dt \tag{291}$$

$$= \frac{8v_0^2}{3D}\left[\frac{1}{2}t^2 - \frac{1}{3}\frac{2v_0}{3D}t^3\right]_0^{3D/2v_0} \tag{292}$$

$$= \frac{8v_0^2}{3D}\left(\frac{1}{2}\left(\frac{3D}{2v_0}\right)^2 - \frac{1}{3}\frac{2v_0}{3D}\left(\frac{3D}{2v_0}\right)^3\right) \tag{293}$$

$$\Delta x = D. \tag{294}$$

c. $a_x(t) = \frac{dv_x}{dt} = \frac{8v_0^2}{3D}\left(1 - \frac{4v_0}{3D}t\right)$. $x(t) = x(0) + \frac{4v_0^2}{3D}\left(t^2 - \frac{4v_0}{9D}t^3\right)$.

Chapter 10. The Concept of Force

1. See Figure 311.

Force diagram for big box

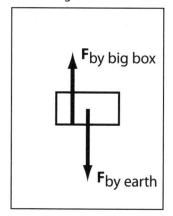

Force diagram for little box

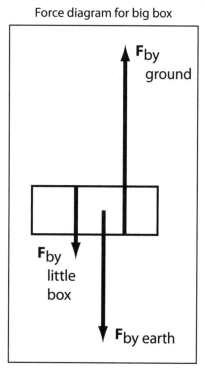

FIGURE 311.
Focused Problem 1.
Force diagrams for
boxes resting atop
one another.

2. See Figure 312. My force diagram for the monkey bar does not include the gravitational force exerted on the monkey bar by the earth. That's because I assumed that the monkey bar is made from a very lightweight material. If that is the case, the gravitational force exerted on the monkey bar by the earth will be negligible compared to the other forces acting on the monkey bar.

Force diagram for orangutan

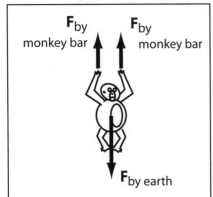

Force diagram for monkey bar

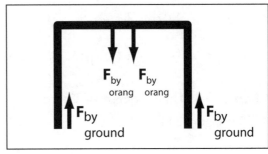

FIGURE 312.
Focused Problem 2.
Force diagrams for
the orangutan and
the monkey bars.

3. See Figure 313. My force diagram for the foot bone does not include the gravitational force exerted on the foot bone by the earth. That's because the weight of the foot bone itself is negligible in comparison with the other forces acting on the foot bone. (All of the other forces acting on the foot bone will be on the order of hundreds of pounds in strength; the foot bone weighs only about a pound or so.)

Take care that your downward force is not labeled $F_{\text{by body}}$. Force requires contact. It is not the person's *body* that touches the foot bone; it is the leg bone that touches the foot bone. Even worse would be labeling this force $F_{\text{by weight}}$, as if it were the person's weight that were pressing down on the foot bone. A weight is not even an *object;* remember that the source of a force must be a *material object* you can point to, such as the leg bone in this case. (Actually, the downward force on the foot bone will be *much larger* than the weight of the person, especially if the person is in the act of leaping off the ground.)

Force diagram for foot bone

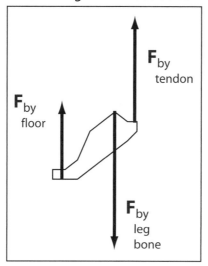

FIGURE 313.
Chapter 10, Focused Problem 3. Force diagram for the foot bone.

Chapter 11. Combining Forces That Act on the Same Target

1. $F_{net} = 500$ N, 37° north of west.

2. $F_D = 25.81$ N, 4.6° below the negative x-axis.

Chapter 12. "Newton's Little Law"

1. Object (a) is speeding up and not turning. Object (b), which has the same force diagram as object (a), is slowing down and not turning. Object (c) is slowing down and turning to its left.

2. Answers follow.
    a. See Figure 314(a).
    b. See Figure 314(b).
    c. See Figure 314(c).
    d. See Figure 315(d).
    e. See Figure 315(e).
    f. See Figure 315(f).
    g. See Figure 315(g).
    h. See Figure 315(h).
    i. See Figure 315(i).

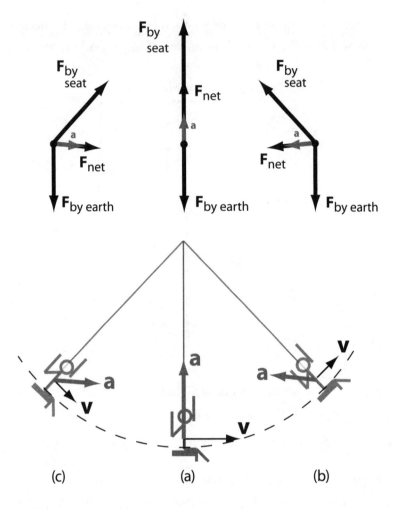

3. Answers follow.

    a. See Figure 316.

    b. The force diagram tells us that the bird's acceleration vector
points straight upward. However, there is no way to know
what the bird's velocity vector looks like at this instant. The
bird could be moving in any direction just now. What we *can*
say is that if the bird is currently moving upward, it must be
in the act of speeding up; if the bird is currently moving
downward, it must be in the act of slowing down; and if
the bird is currently moving left or right, it must be at the
bottom of a dive, turning upward.

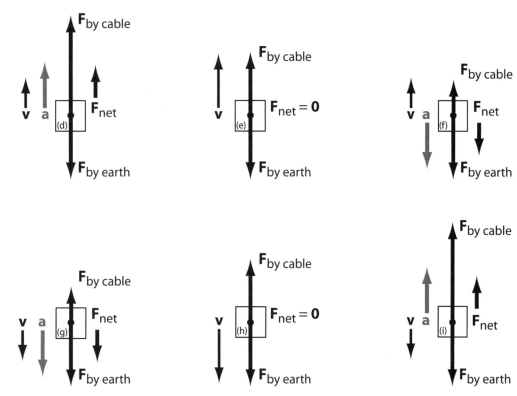

FIGURE 315. Focused Problem 2(d)–(i). Force diagrams for an elevator in motion. (Velocity and acceleration vectors reproduced from Figure 284.)

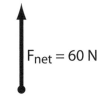

FIGURE 316. Focused Problem 3. Net force acting on a bird at a particular moment of time.

   c. The bird could even be at rest at this particular moment. In that case, the bird would be just *about* to move upward. Perhaps this bird fell out of its nest and has been flapping its wings to generate a steady 60 N worth of lift; the bird's efforts have finally succeeded in bringing it to rest, and in the next moment it will be moving back upward.

4. Figure 317 illustrates the following scenarios:
   a. A baseball player sliding into second base.
   b. A person riding over the top of a Ferris wheel.

c. A baby being lifted out of its crib at a constant speed.

d. A rocket after liftoff, gaining speed.

e. A bungee jumper at the bottom of her downward motion, about to slingshot back upward.

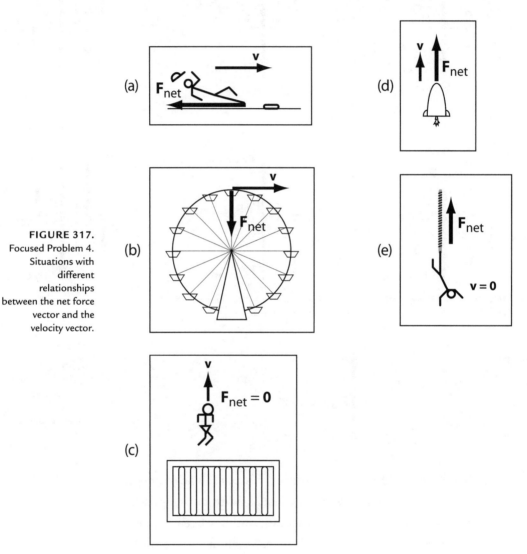

FIGURE 317.
Focused Problem 4.
Situations with
different
relationships
between the net force
vector and the
velocity vector.

Only in (b) is the trajectory curved at the moment in question. This is because only in (b) does $\mathbf{F}_{net}$ (and hence **a**) have a turning component (that is, a component perpendicular to **v**).

The object is speeding up in (d), slowing down in (a), and neither speeding up nor slowing down in (b), (c), and (e).

In every case, $\mathbf{F}_{net}$ and $\mathbf{a}$ point the same way.

5. See Figure 318. Because the speed is constant in my theory, $a_{\parallel} = 0$ and $\mathbf{a}$ is perpendicular to the velocity vector. Hence, $\mathbf{F}_{net}$ must also be perpendicular to the velocity vector, as shown in the figure.

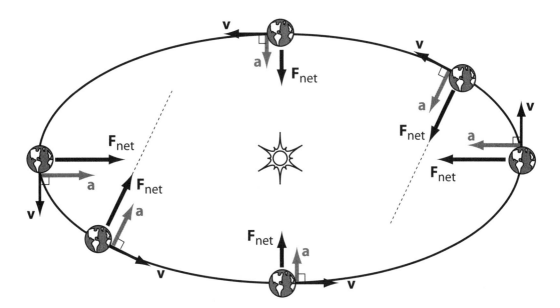

FIGURE 318.
Focused Problem 5.
An unsatisfactory
theory of gravity.

The main flaw in my theory is that the force that the sun exerts on the earth isn't always pointing directly toward the sun—which doesn't seem sensible. (Note the dashed lines in Figure 318.)

In the following chapters we'll see that $\mathbf{F}_{net}$ is larger or smaller according to whether the acceleration vector is larger or smaller. I made Figure 318 reflect this fact. (The acceleration vectors have magnitude $a_{\perp} = v^2/r$; the speed stays constant, but the radius of curvature varies from place to place. Therefore, the magnitude of $\mathbf{a}$ varies from place to place, so $\mathbf{F}_{net}$ varies from place to place.) Now we see another strange feature of my theory: The force exerted on the earth by the sun is *stronger* when the earth is *farther away!* This doesn't seem sensible, either.

6. Answers follow.

   a. See Figure 319.

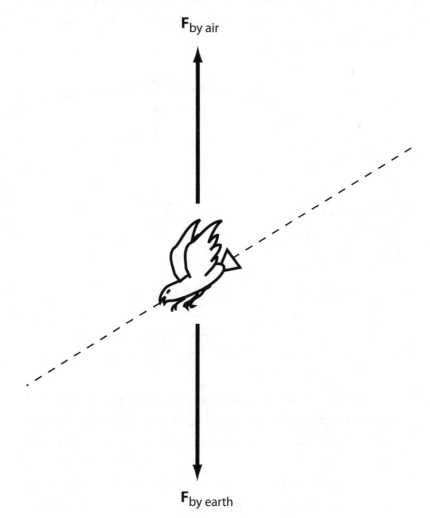

F_{by air}

FIGURE 319.
Focused Problem 6.
Force diagram for
the hawk.

F_{by earth}

   b. I don't know much about avian flight, but I do know New-ton's Little Law. The hawk is not speeding up, not slowing down, and not turning, so $\mathbf{a} = 0$; hence $\mathbf{F}_{net} = 0$. We know that the force on the hawk from the earth points straight down. So we need $\mathbf{F}_{by\ air}$ to point straight up in order to cancel $\mathbf{F}_{by\ earth}$. If there were any horizontal component in

$F_{by\ air}$, there would necessarily be a horizontal component in F_{net}. This, in turn, would give **a** a horizontal component, making the hawk turn as well as speed up or slow down.

c. Not only must $F_{by\ air}$ point straight up, but the magnitude of $F_{by\ air}$ must be equal to the magnitude of $F_{by\ earth}$. Otherwise, the two forces won't cancel completely, and there will be a leftover vertical component to F_{net}. This, in turn, will give a vertical component to **a**, making the hawk turn as well as speed up or slow down.

If we wish to think about the details, we can subdivide the force from the air into two main contributions. First of all, there is going to be some air resistance, or drag, pushing the hawk backward (opposite to the velocity vector). Second, the hawk will adjust its wings so that the air provides a lift force, pointing up and a little forward. See Figure 320. These two forces from the air must combine to generate an upward force that is equal and opposite to the force from the earth. Otherwise we will not observe the trajectory described in the problem.

7. Answers follow.

a. See Figure 321.

b. I don't know all that much about helicopters, but I do know Newton's Little Law. The helicopter is not speeding up, not slowing down, and not turning, so **a** = 0; hence $F_{net} = 0$. We know that $F_{by\ earth}$ points straight down. So we need $F_{by\ air}$ to point straight up in order to cancel the force from the earth. If there were any horizontal component in $F_{by\ air}$, there would necessarily be a horizontal component in F_{net}. This, in turn, would give a horizontal component to **a**, making the helicopter turn as well as speed up or slow down.

c. Not only must $F_{by\ air}$ point straight up, but the magnitude of $F_{by\ air}$ must be equal to the magnitude of $F_{by\ earth}$. Otherwise, the two forces won't cancel completely and there will be a leftover vertical component to F_{net}. This, in turn, will give a

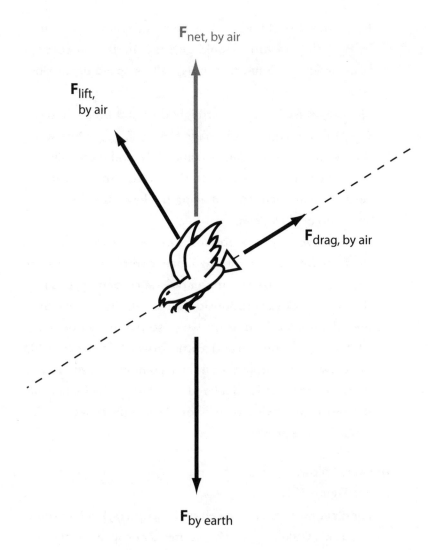

FIGURE 320.
Focused Problem 6.
Examining $F_{by\ air}$
more closely. Note
that $F_{lift,\ by\ air}$ +
$F_{drag,\ by\ air}$ = $F_{net,\ by\ air}$.

vertical component to **a**, making the helicopter turn as well as speed up or slow down.

If we wish to think about the details, we can subdivide the force from the air into two main contributions. First of all, there is going to be some air resistance, or drag, pushing the helicopter backward (opposite to the velocity vector). Second, the helicopter pilot will adjust the controls so that the air provides a lift force, pointing up and a little forward. See Figure 322. These two forces from the air must combine to generate an upward force that is equal and opposite to

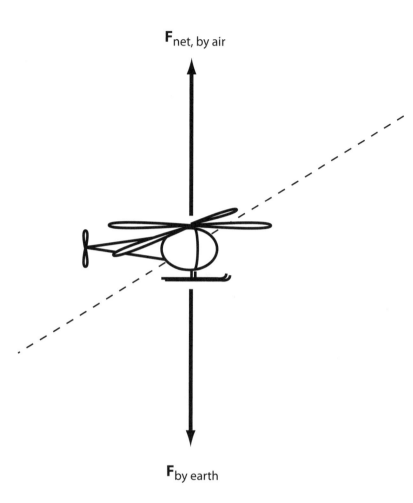

$\mathbf{F}_{\text{net, by air}}$

$\mathbf{F}_{\text{by earth}}$

FIGURE 321.
Focused Problem 7.
Force diagram for the
helicopter.

the force from the earth. Otherwise we will not observe the
trajectory described in the problem.

8. See Figure 323. I am assuming that the shopper is not speeding up, not slowing down, and not turning ($\mathbf{a} = \mathbf{0}$, so we must
have $\mathbf{F}_{\text{net}} = \mathbf{0}$). In the diagram to the left, air resistance has
been neglected. In the diagram to the right, air resistance
has been included. (The magnitude of the air resistance has
been exaggerated for the sake of clarity.) Including air resistance has two effects. First, in order for the shopper to maintain a constant speed along a straight-line trajectory, there
must be a horizontal force from the step; this force is frictional
in nature. We see that it would be impossible to ride a perfectly
frictionless escalator; air resistance would hold you in place as

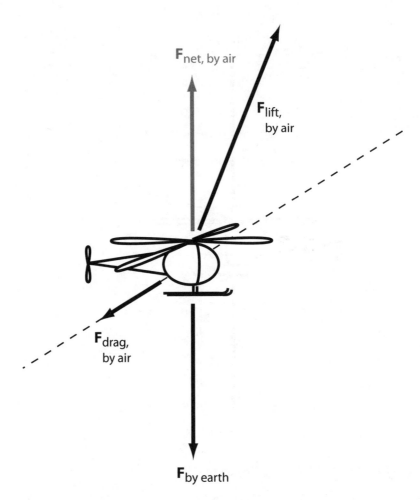

FIGURE 322.
Chapter 12,
Focused Problem 7.
Examining $\mathbf{F}_{\text{by air}}$
more closely. Note
that $\mathbf{F}_{\text{lift, by air}} +$
$\mathbf{F}_{\text{drag, by air}} = \mathbf{F}_{\text{net, by air}}$.

the steps slip past beneath your feet! Second, because the air
resistance presses you down slightly, the upward vertical force
from the step must increase somewhat; otherwise the net force
would be downward, meaning that your trajectory would curve
downward and you would sink into the step!

Chapter 13. Newton's Second Law

1. Answers follow.
    a. 3.33 m/s^2.
    b. $37°$ north of west.
    c. The information in the problem gives us no way to know the
        speed or the direction of motion of the cart at this particular
        instant.

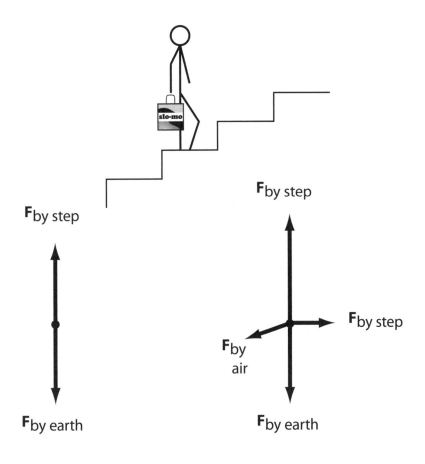

FIGURE 323.
Chapter 12, Focused Problem 8. Force diagrams for a shopper riding up an escalator at a constant speed, in a straight line. The left-hand diagram neglects air resistance; the right-hand diagram includes an exaggerated air resistance.

2. Answers follow.
   a. See Figure 324.
   b. See Table 8.

3. With an acceleration of 90 N/7 kg = 12.86 m/s² for a period of 1 s, the maximum speed reached is 12.86 m/s. The distance traveled during $t_1 \rightarrow t_2$ is then computed as 6.43 m. The distance traveled during $t_2 \rightarrow t_3$ is 25.7 m. The acceleration during the interval $t_3 \rightarrow t_4$ is 10 N/7 kg = 1.43 m/s². This determines the duration of the interval as $\left(0\frac{m}{s} - 12.86\frac{m}{s}\right)/\Delta t = -1.43$ m/s² $\Rightarrow \Delta t = 9$ s. The distance traveled during $t_3 \rightarrow t_4$ is then computed as 57.9 m. Total distance traveled: 90 m.

4. Answers follow.
   a. See Figure 325.

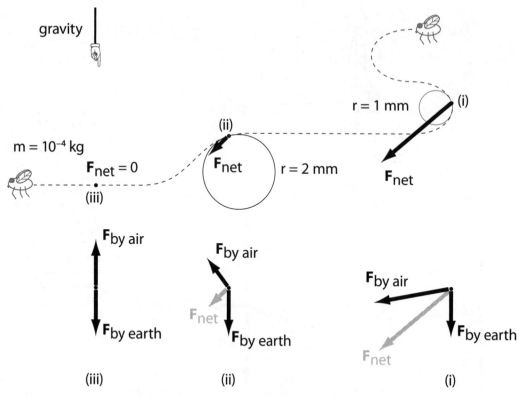

FIGURE 324. Focused Problem 2(a). Force diagrams for a fly at different moments in its motion.

TABLE 8. Focused Problem 2(b)

Moment	v	$a_{\parallel}$	r	$a_{\perp}$	a	F_{net}
(i)	0.1 m/s	0.05 m/s^2	1 mm	10 m/s^2	10.00001 m/s^2	0.001 N
(ii)	0.0112 m/s	0.05 m/s^2	2 mm	0.06245 m/s^2	0.08 m/s^2	8×10^{-6} N
(iii)	1 m/s	0	∞	0	0	0

b. At all times during the clown's flight, the acceleration vector **a** points directly downward (in the negative y-direction). The direction of **a** does not change over time. The magnitude of **a** does not change over time. The magnitude is $a = F_{net}/m = (980\text{ N})/(100\text{ kg}) = 9.8\text{ m/s}^2$.

c. $a_x = 0$, $a_y = -9.8\text{ m/s}^2$.

d. $\theta_0 = \arctan(40/20) = 63.4°$. $v_0 = \sqrt{v_{0,x}^2 + v_{0,y}^2} = 44.7\text{ m/s}$.

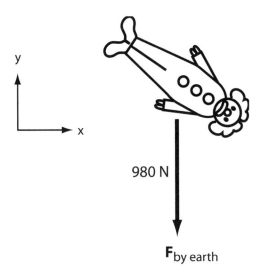

FIGURE 325.
Focused Problem 4.
Force diagram for
a clown during his
flight.

y

x

980 N

$\mathbf{F}_{\text{by earth}}$

e. Because $a_x = 0$, v_x is constant over time, with the value $v_x = +20$ m/s. Thus $v_{\text{peak},x} = +20$ m/s. Meanwhile, $v_{\text{peak},y} = 0$ because the clown is traveling purely horizontally at this moment. His peak speed is $v_{\text{peak}} = \sqrt{v_{\text{peak},x}^2 + v_{\text{peak},y}^2} = 20$ m/s.

f. Set $v_y = 0$ in $v_y(t) = (+40 \text{ m/s}) + (-9.8 \text{ m/s}^2)t$ to find $t_{\text{peak}} = 4.08$ s. Now $y_{\text{peak}} = (0) + (+40 \text{ m/s})t_{\text{peak}} + \frac{1}{2}(-9.8 \text{ m/s}^2)t_{\text{peak}}^2 = 81.6$ m.

g. Set $y = 0$ in $y(t) = 0 + (+40 \text{ m/s})t + \frac{1}{2}(-9.8 \text{ m/s}^2)t^2$ to find $t_{\text{land}} = 8.16$ s. Now $x_{\text{land}} = 0 + (+20 \text{ m/s})t_{\text{land}} + (0)t_{\text{land}}^2 = 163.2$ m.

5. Answers follow.

   a. See the graph of $F_{\text{net},y}$ in Figure 326. We are taking the positive y-direction to be upward.

   b. See the lower right-hand graph in Figure 326.

   c. From the force diagrams in Figure 315 we have $F_{\text{by cable},y} + F_{\text{by earth},y} = ma_y$. And $F_{\text{by earth},y}$ is given to be $-5{,}000$ N. So $F_{\text{by cable},y} = ma_y + 5{,}000$ N. And from the information given, we can find a_y during the acceleration phases: $a_y = \frac{\Delta v_y}{\Delta t} = \frac{\pm 2 \text{ m/s}}{0.5 \text{ s}} = \pm 4$ m/s^2. Using $m = 500$ kg, we therefore have $F_{\text{net},y} = ma_y = (500 \text{ kg})(\pm 4 \text{ m/s}^2) = \pm 2{,}000$ N during the acceleration phases. This leads to $F_{\text{by cable},y} = \pm 2{,}000$ N + $5{,}000$ N, as shown at lower right in Figure 326.

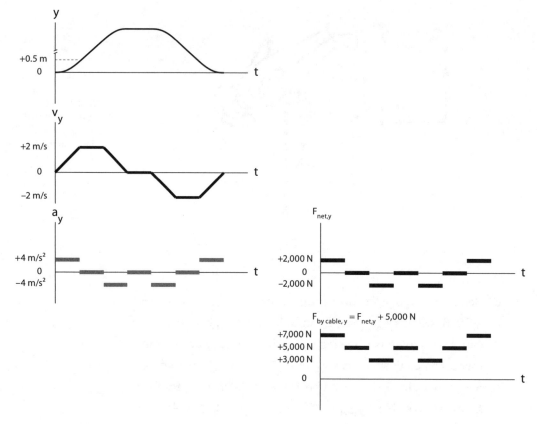

FIGURE 326.
Focused Problem 5.
$y - t, v_y - t, a_y - t,$
and $F_{net,y} - t$ graphs
for an elevator.

d. Based on the graph of $F_{by\ cable,y}$, the cable is most likely to break when the tension in the cable is 7,000 N, that is, when the elevator is being hauled up from the ground and when it is being lowered back to the ground. You can feel this in your bones if you lift a heavy bucket to emulate the motion of the elevator.

Chapter 14. Dynamics

1. See Figure 327. The floor is exerting this force on the person.

2. See Figure 328. The operator of the trolley car is applying the brakes more and more strongly as the car comes to a stop. This is the kind of stop that makes you continually modify your stance and grab the handrails more and more strongly as the

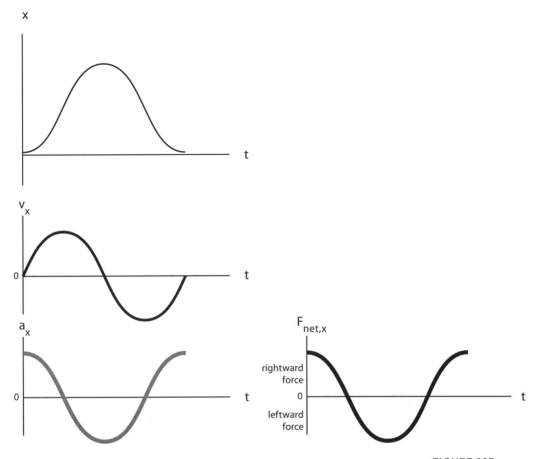

FIGURE 327.
Focused Problem 1.
$x - t$, $v_x - t$, $a_x - t$,
and $F_{net,x} - t$ graphs
for a person pacing
the floor.

process unfolds. It's the kind of stop that lurches hardest at the very end.

3. $v_x(t) = 10 \text{ m/s} - (\frac{2}{5}\frac{m}{s^3})t^2$. $x(t) = x(0) + (10 \text{ m/s})t - (\frac{2}{15}\frac{m}{s^3})t^3$. $x(5 \text{ s}) - x(0) = 33\frac{1}{3}$ m.

4. Answers follow.

   a. When the pelican releases the fish, the fish starts off with the same velocity the pelican had, $v_0 = (+5 \text{ m/s}, 0)$. I take the origin as the location of the otter. Set $y = 0$ in $y(t) = (+20 \text{ m}) + \frac{1}{2}(-9.8 \text{ m/s}^2)t^2$ to find $t_{splash} = 2.02$ s. Now $x_{splash} = (0) + (+5 \text{ m/s})t_{splash} + (0)t_{splash}^2 = 10.1$ m. The fish lands 10.1 meters from the otter.

   b. Because $a_x = 0$, v_x remains constant at 5 m/s during the flight. To find v_y at splashdown, put $t_{splash} = 2.02$ s into

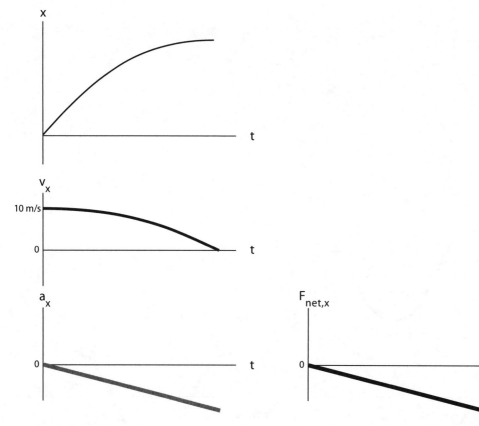

FIGURE 328.
Focused Problem 2.
$x - t, v_x - t, a_x - t,$
and $F_{net,x} - t$ graphs
for a trolley car.

$v_y(t) = (0) + (-9.8 \text{ m/s}^2)t$ to find $v_y = -19.8$ m/s. The fish strikes the water with speed $v = \sqrt{v_x^2 + v_y^2} = 20.4$ m/s, moving in the direction $\theta = \arctan(19.8/5) = 76°$ below the horizontal.

5. Answers follow.

    a. I take the origin as the release point. The launch velocity has components $v_{0,x} = +(10 \text{ m/s}) \cos 30° = +8.66$ m/s and $v_{0,y} = +(10 \text{ m/s}) \sin 30° = +5$ m/s. Set $x = 10$ m in $x(t) = (0) + (+8.66 \text{ m/s})t + \frac{1}{2}(0)t^2$ to find $t_{strike} = 1.155$ s. Now $y_{strike} = (0) + (+5 \text{ m/s})t_{strike} + \frac{1}{2}(-9.8 \text{ m/s}^2)t_{strike}^2 = -0.76$ m. The ball strikes the wall at a point 0.76 vertical meters below the release point, or 1.5 m $-$ 0.76 m $= 0.74$ m above the ground.

b. Because $a_x = 0$, v_x remains constant at 8.66 m/s during the flight. To find v_y at impact, put $t_{strike} = 1.155$ s into $v_y(t) = (+5$ m/s$) + (-9.8$ m/s$^2)t$ to find $v_y = -6.3$ m/s. The ball strikes the wall with speed $v = \sqrt{v_x^2 + v_y^2} = 10.7$ m/s, moving in the direction $\theta = \arctan(6.3/8.66) = 36°$ below the horizontal. The ball is on its way down when it hits ($v_y < 0$).

6. I take the origin as the location of the volcano. Set $v_y = 0$ in $v_y(t) = v_0 \sin 45° + (-9.8$ m/s$^2)t$ to find $t_{peak} = \left(0.072\frac{s^2}{m}\right) v_0$. Now put $t = t_{peak}$ and $y = 20$ mi $= 32{,}200$ m in $y(t) = (0) + (v_0 \sin 45°)t + \frac{1}{2}(-9.8$ m/s$^2)t^2$ to find $\left(0.036\frac{s^2}{m}\right)v_0^2 = 32{,}200$ m, or $v_0 = 950$ m/s $= 2{,}100$ mph! (If you rounded numbers differently than I did, your answer might be somewhat different, but in any case your launch speed should be in the neighborhood of 1,000 m/s or so.)

7. Answers follow.
    a. $\frac{d^2x}{dt^2} = -\frac{k}{m}x$.
    b. With $x(t) = x_0 \cos \omega t + \frac{v_{0,x}}{\omega} \sin \omega t$, we will have $\frac{d^2x}{dt^2} = -\omega^2(x_0 \cos \omega t + \frac{v_{0,x}}{\omega} \sin \omega t) = -\omega^2 x$, which satisfies the Second Law provided that $\omega = \sqrt{\frac{k}{m}}$. Note that we will also have $x(0) = x_0 \cos 0 + \frac{v_{0,x}}{\omega} \sin 0 = x_0$ and $v_x(0) = \frac{dx}{dt}\big|_{t=0} = (-\omega x_0 \sin \omega t + v_{0,x} \cos \omega t)_{t=0} = v_{0,x}$ as required by the initial conditions.
    c. See Figure 329.

Chapter 15. Newton's Third Law

1. See Figure 330. The horse has neglected to mention the force that the *ground* exerts on him. If the magnitude of $F_{by\ ground}$ is greater than the magnitude of $F_{by\ cart}$, F_{net} will point to the right, and this will give the horse a nonzero acceleration vector pointing to the right, which means that the horse will speed up from rest and begin moving to the right.

2. The bully probably contains more mass than the nerd, so it can be tempting to think of this situation as being similar to that in Worked Example 2 of this chapter, the collision between the cement truck and the motorcycle. But the mass disparity is

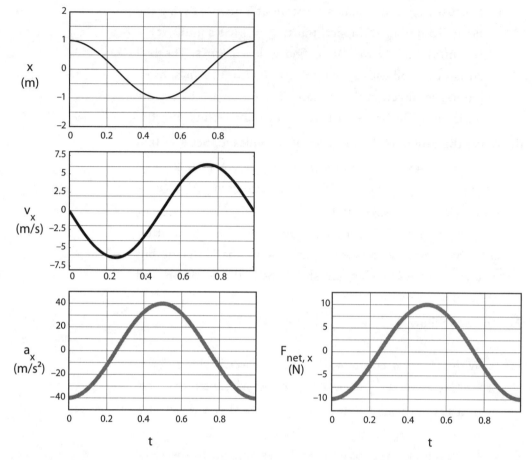

FIGURE 329.
Chapter 14, Focused
Problem 7. $x - t$,
$v_x - t$, $a_x - t$, and
$F_{net,x} - t$ graphs for a
mass on a spring.

not nearly so great in this case. The real reason the nerd goes flying while the bully stays still is that the bully *sets his feet* before shoving the nerd. See Figure 331.

The bully's posture allows the ground to push him to the right during the encounter, which cancels the leftward push from the nerd, keeping the bully stationary. Meanwhile, the nerd has not set his feet, so he enjoys no stabilizing effect from the ground.

3. In this case, the nerd has learned the value of surprise. Now the *nerd* sets his feet before shoving the bully. The forces between the two are equal during the interaction, as required by the Third Law, but now it is the nerd who enjoys the stabilizing

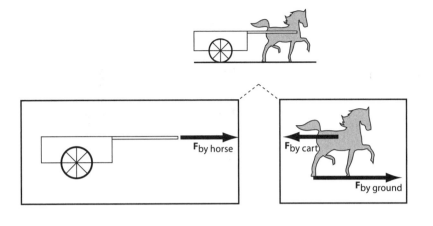

FIGURE 330.
Focused Problem 1. Force diagrams for the horse and the cart. Only horizontal forces are shown.

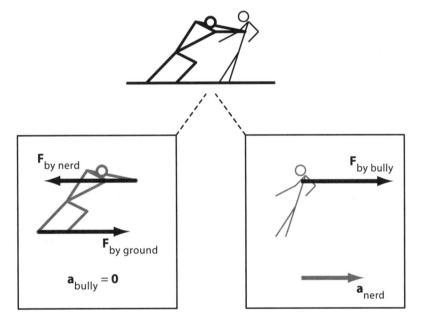

FIGURE 331.
Focused Problem 2. Force diagrams for the bully and the nerd. Only horizontal forces are shown.

effect of a force from the ground. The force diagrams are like those shown in Figure 331, but with the roles of the two children reversed.

4. See Figure 332 (and compare it to Figure 312). Forces 1 and 3 make up a Third Law pair. Forces 2 and 4 make up a Third Law pair.

5. The hummingbird cannot lift the box this way, no matter how strong she is and no matter how light the box is. Intuitively, the trouble is that, as hard as she presses up on the box, she

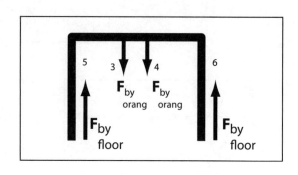

FIGURE 332.
Focused Problem 4.
Force diagrams for
the orangutan and
the monkey bars.

presses down on the air. The air transmits this downward push to the box, so the net effect on the box cancels.

To approach the problem more carefully, let us entirely neglect gravitational forces and draw force diagrams as in Figure 333.

Considering Third Law pairs, we have $F_1 = F_3$, $F_2 = F_5$, and $F_4 = F_6$. In order for the hummingbird to get moving from rest, F_1 must exceed F_2, as suggested in the leftmost force diagram. And in order for the air to get moving from rest, F_4 must exceed F_3, as suggested in the middle force diagram. So putting things together, in order for the hummingbird to lift the box we must have

$$F_6 = F_4 \tag{295}$$
$$> F_3 \tag{296}$$
$$= F_1 \tag{297}$$
$$> F_2 \tag{298}$$
$$= F_5, \tag{299}$$

that is, $F_6 > F_5$, as suggested by the rightmost force diagram. But this says that the net force on the box would be *downward*, not upward. Hence the hummingbird cannot lift the box. The best she can do is remain stationary, in which case all of the forces in Figure 333 will be equal in magnitude.

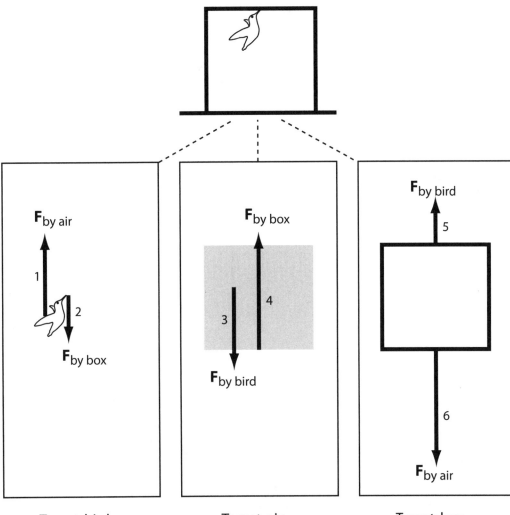

Target: bird   Target: air   Target: box

FIGURE 333.
Chapter 15, Focused
Problem 5. Force
diagrams for the
hummingbird, the
air in the box, and the
box itself.

Chapter 16. Kinds of Force

1. See Figure 334. The significant forces acting on the car are gravitational, normal, and frictional in nature. The friction force is static and points uphill. You can add the five vectors tip-to-tail to see that $\mathbf{F}_{net} = 0$.

2. See Figure 335. The forces acting on the car are gravitational, normal, and frictional in nature. The friction force is static and points uphill. You can add the five vectors tip-to-tail to see that $\mathbf{F}_{net} = 0$.

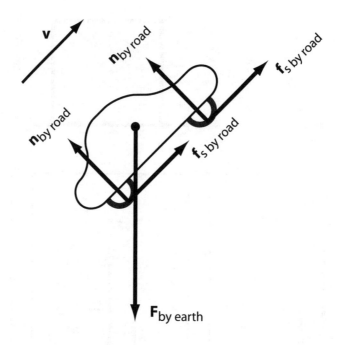

FIGURE 334.
Focused Problem 1.
Force diagram for a
car driving up a
straight hill at
constant speed.

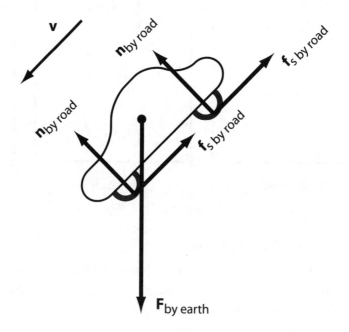

FIGURE 335.
Focused Problem 2.
Force diagram for a
car driving down a
straight hill at
constant speed.

3. The forces on the car are exactly the same in the two situations.
    By the way, here's how I originally devised this problem. When
    I commute to work every morning, I drive up a steep hill. On
    my way home at night, I drive down that same hill. One day I

realized in my bones that the forces on me were the same in the two situations. In the morning, when I am driving up the hill, the back of the car seat is pushing me uphill; I feel it in my shoulders. At night, when I am driving down the hill, the steering wheel is pushing me uphill; I feel it in my wrists. The uphill force is the same in each case; it's just that a different part of the car provides the force in the two situations! (And a different part of my body feels the force in the two situations.)

4. Because the car is moving on a straight, flat, road, there is no turning, so $a_\perp = 0$, that is, $a_y = 0$. This means that forces in the vertical direction cancel each other, so the total normal force from the ground must have the value $n = mg$.

   Meanwhile, because the car is only gradually speeding up, we take $a_\parallel \approx 0$ as well, that is, $a_x = 0$. This means that forces in the horizontal direction cancel each other, so the static friction force must have the value $f_s = hv^2$. As the car's speed increases, hv^2 will eventually exceed the value $f_s^{\text{break}} = \mu n = \mu mg$. So the wheels will begin to slip when $hv^2 = \mu_s mg$, or when $v = \sqrt{\mu_s mg/h}$.

Chapter 17. Strategies for Applying Newton's Laws

A force diagram is provided for each of the problems in the chapter. This is usually the first thing to make sure you have right.

1. Answers follow.
   a. $f_s = 0.2$ N. See Figure 336.
   b. $f_s = 0$. See Figure 337.

2. $r_{\text{min}} = 28.7$ m. See Figure 338.

3. $a_x = 1.1\,\text{m/s}^2$, $a_y = 0$. See Figure 339.

4. Answers follow.
   a. See Figure 340.
   b. See Figure 341. Pulling at too steep an angle is inefficient because you are exerting a lot of upward force to cancel gravity when the ground could be doing more of that work for you. But pulling at too shallow an angle is also inefficient because you miss an opportunity to decrease friction by

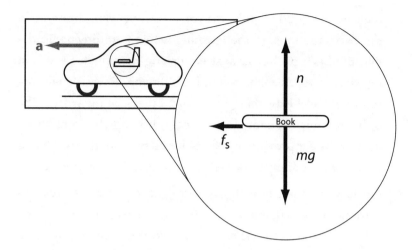

FIGURE 336.
Force diagram for
Focused
Problem 1(a).

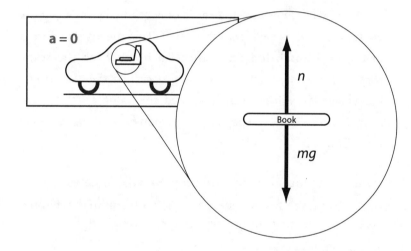

FIGURE 337.
Force diagram for
Focused
Problem 1(b).

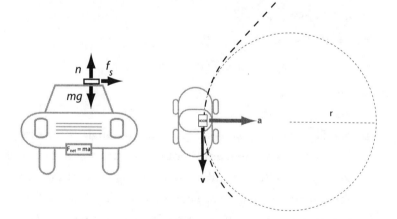

FIGURE 338.
Force diagram for
Focused Problem 2.

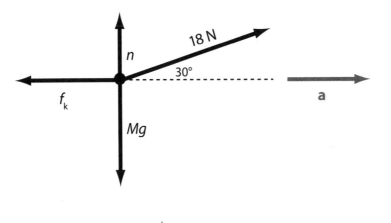

FIGURE 339.
Force diagram for
Focused Problem 3.

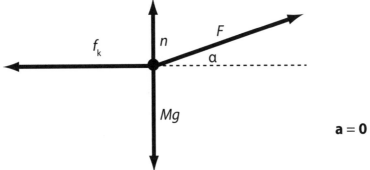

FIGURE 340.
Force diagram for
Focused Problem 4.

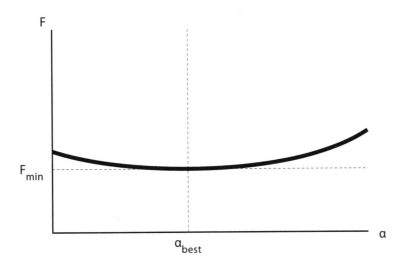

FIGURE 341.
A graph of the
required pulling force
F versus the pulling
angle α for the case
$\mu_k = 0.8$.

lifting up slightly (which decreases the normal force and hence decreases the kinetic friction force).

c. The function $F(\alpha) = Mg \left(\frac{\mu_k}{\cos \alpha + \mu_k \sin \alpha} \right)$ can be minimized by maximizing the denominator. Setting the derivative of the denominator equal to zero gives $\tan \alpha = \mu_k$. Thus $\alpha_{best} = \tan^{-1} \mu_k$. Putting α_{best} back into $F(\alpha)$, after a little simplification we obtain $F_{min} = \frac{\mu_k}{\sqrt{1 + \mu_k^2}} Mg$.

5. See Figure 342.

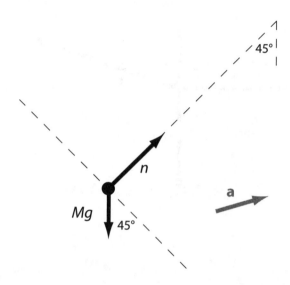

FIGURE 342.
Force diagram for Focused Problem 5.

a. $a_\perp = 16.2 \text{ m/s}^2$.
b. $a_\parallel = 6.9 \text{ m/s}^2$.
c. $n = 2{,}310 \text{ N}$.

6. Answers follow.
   a. See Figure 343. The scale reads 784 N.
   b. See Figure 344. The scale reads 781.3 N. This is lower than your actual weight by 2.7 N, about 10 ounces in ordinary terms. ("I lost ten ounces in one day by moving to Ecuador!")
   c. Put $n = 0$ to find that if the day were roughly one hour and 24 minutes long, your feet would just graze the scale at the equator. (Likewise, rocks, gravel, and animals would hover along the ground.)

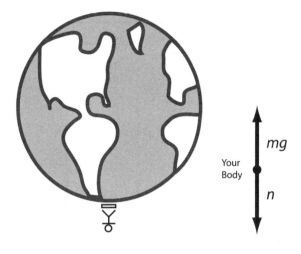

FIGURE 343.
Force diagram
for Focused
Problem 6(a).

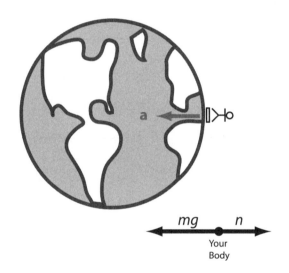

FIGURE 344.
Force diagram
for Focused
Problem 6(b).

7. Answers follow.
   a. $a = \left(\frac{m - \mu_k M}{m + M}\right) g = 1.39$ m/s². See Figure 345.
   b. $a = \left(\frac{\sin\theta - \mu_k \cos\theta - \frac{m}{M}}{1 + \frac{m}{M}}\right) g = 2.24$ m/s². The large block acceler-
      ates downhill for the given values. See Figure 346.

8. $m = M(\mu_s \cos 15° + \sin 15°) = 18.7$ kg. See Figure 347.

9. Answers follow.
   a. See Figure 348.
   b. Put $n = 0$ to find $v_{\text{float}} = \sqrt{gR} = 7$ m/s.

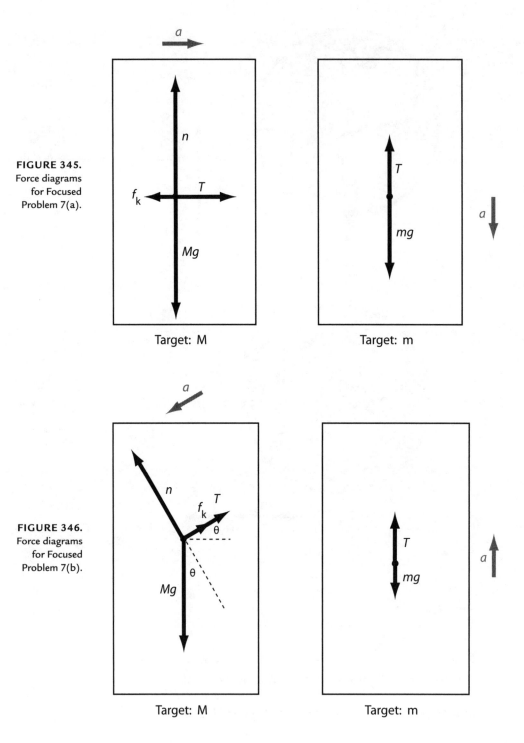

FIGURE 345.
Force diagrams
for Focused
Problem 7(a).

a

n

f_k T

Mg

Target: M

T

mg

a

Target: m

FIGURE 346.
Force diagrams
for Focused
Problem 7(b).

a

n

f_k T

θ

θ

Mg

Target: M

T

mg

a

Target: m

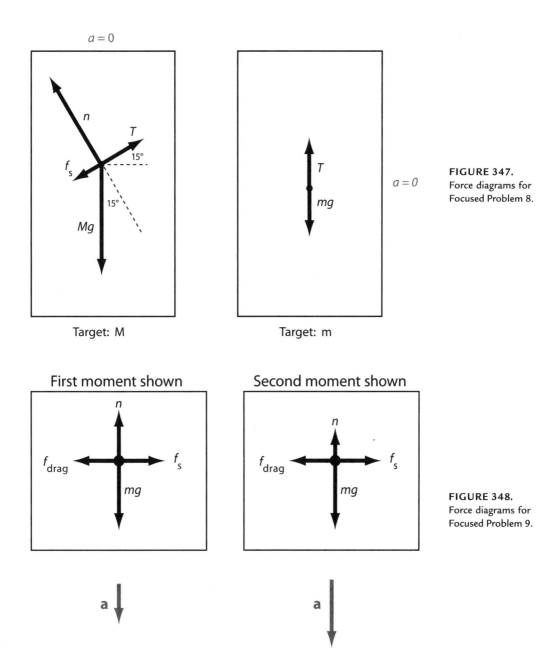

$a = 0$

n

T

15°

f_{s}

15°

Mg

Target: M

T

mg

$a = 0$

Target: m

FIGURE 347.
Force diagrams for
Focused Problem 8.

First moment shown

n

f_{drag} f_{s}

mg

a

Second moment shown

n

f_{drag} f_{s}

mg

a

FIGURE 348.
Force diagrams for
Focused Problem 9.

10. Answers follow.

   a. See Figure 349. The Second Law gives

(300)
$$k(L - D) = \frac{GM^2}{D^2}.$$

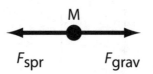

   b. With values of k, L, G, and M specified, Equation 300 deter-
      mines the value(s) of D for which the star is in equilibrium.

      A useful technique is to graph the left-hand side of
      Equation 300 and the right-hand side of Equation 300 on
      the same set of axes. See Figure 350.

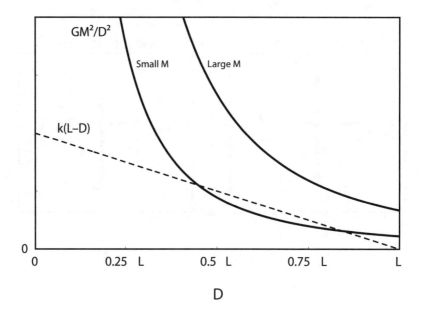

      For a large stellar mass M, the curves do not intersect.
      Therefore, equilibrium is impossible and the star must col-
      lapse. For a small stellar mass M, the curves intersect at two
      points, so there are two equilibrium diameters for the star.

      To analyze the situation in more detail, rewrite Equa-
      tion 300 in the form $z^2(1 - z) = \lambda$, where $z \equiv D/L$ and
      $\lambda \equiv GM^2/kL^3$. For z in $[0, 1]$ the left-hand side has the

maximum value 4/27, so when $\lambda > 4/27$ there is no so-lution. This shows that equilibrium is impossible when $M > \sqrt{4kL^3/27G}$. If we put $M = bL^3$, where b is a typical stellar density, the collapse condition becomes $M > \frac{4}{27}\frac{k}{bG}$.

11. Answers follow.

a. See Figure 351.

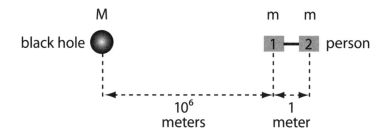

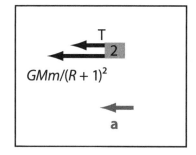

Target: block 1   Target: block 2

FIGURE 351.
Focused Problem 11.
Force diagrams for
the two halves of the
astronaut.

b. $T = \frac{1}{2}GMm\left(\frac{1}{R^2} - \frac{1}{(R+1)^2}\right) = 13{,}340\,\text{N}$, or about 3,000 pounds of tension in the spine.

c. The closer the person gets to the black hole, the greater the tension will become, and the person will be pulled apart like taffy. Eventually the person will be stretched into a long chain of molecules, then a long chain of atoms, then a long chain of nuclei. See Figure 352.

FIGURE 352. Focused Problem 11. Why it doesn't pay to get too close to a black hole.

References

[1] United Nations Intergovernmental Panel on Climate Change, Working Group 1. "Changes in Atmospheric Constituents and Radiative Forcing." Chapter 2 in The Physical Basis of Climate Change. Assessment Report 4, FAQ 2.1, Figure 1, page 135, 2007. Available at www.ipcc.ch/pdf/assessment-report/ar4/wg1/ar4-wg1-chapter2.pdf.

[2] Bressoud, D. M. "The Fundamental Theorem of Calculus," 2005. Available at www.macalester.edu/~bressoud/talks/.

[3] Giesen, J. Javascript computation of sunrise and sunset, 2007. Available at www.jgiesen.de/astro/astroJS/decEoT/index.htm.

[4] Cohen, I. B., ed. Isaac Newton's Papers and Letters on Natural Philosophy, and Related Documents. Edited and with a general introduction by I. Bernard Cohen, assisted by Robert Schofield. Cambridge: Cambridge University Press, 1958.

[5] Zimba, J. "Inertia and Determinism." British Journal for the Philosophy of Science 59, no. 3 (2008): 417–428.
    In this article, which was actually inspired by the writing of this book, I raise—and attempt to clarify—some intricacies in the relationship between the First and Second Laws.

[6] Laplace, P.-S. A Philosophical Essay on Probabilities. Trans. F. W. Truscott and F. L. Emory. New York: Wiley, 1902.

[7] Winchester, S. Krakatoa: The Day the World Exploded, August 27, 1883. New York: HarperCollins, 2003.

[8] Weihs, D. "The Hydrodynamics of Dolphin Drafting." Journal of Biology 3 (2004): 801–816.

Index of Problem Situations

This index presents a list of situations analyzed in the Worked Examples and Focused Problems.

Subject Index

accelerating reference frame, 257–61, 296, 332
acceleration: component parallel to velocity
   ($a_\parallel$), 95–97, 104, 112, 120, 133; component
   perpendicular to velocity ($a_\perp$), 95–97, 127–
   33, 251, 322, 328, 339–45; computing
   from velocity, 104–8, 152–53; concept map
   for, 133; constant, 159–75; coordinate
   components of (a_x and a_y), 108–20, 133;
   Huygens's Formula for, 127–33, 339–45;
   meaning of, 87, 133, 234; negative, 159; in
   one-dimensional problems, 141–42; as rate
   of change of velocity, 99, 103–9, 111–20,
   128, 133, 142, 152–53, 156, 158–59, 254;
   sketching vector for, 90–92; slowing-down
   part, 95–97, 104, 112, 120, 133; speeding-
   up part, 95–97, 104, 112, 120, 133; while
   standing still, 98–99, 153; as "turn signal,"
   92, 96; turning part, 95–97, 127–33, 251,
   322, 328, 339–45
acceleration-time graphs, 148–50; from force-
   time graphs, 269; observations about, 154;
   from velocity-time graphs, 152–53
action at a distance, 202–4
action-reaction, 283–87
air resistance, 293, 311, 312–33
Aristotle, 201
atoms, 277–78

bathroom scale, 300
brains, 277
breaking point, 306–7

Cauchy, Augustin, 23
cheekbones, 259–60
circular motion, 127–29, 339–41
climate change, 3
coasting, 97, 147
collision problems, 175–79
combining forces, 44, 215–26, 300, 323–24
components, 53–58, 64–66, 110–11
constant-acceleration problems: acting out,
   165–66; formulas for, 167, 187, 189, 270;
   sketching solution, 160–5
coordinate systems: importance of, 51, 148–49,
   270; origin in, 33–34; tilted, 52
curvature: identifying visually, 88–90; radius of,
   129–30, 341–44

dancer, 166
deceleration, 87
Depp, Johnny, 260
Descartes, René, 23
determinism, 160, 276–78. *See also* mechanistic
   universe
direction, 62–63

About the Author

Jason Zimba is a faculty member in physics and mathematics at Bennington College and has taught at Grinnell College and the University of California, Berkeley. He was the recipient in 2006 of the Majorana Prize.